# SpringerBriefs in Physics

SpringerBriefs in Physics are a series of slim high-quality publications encompassing the entire spectrum of physics. Manuscripts for SpringerBriefs in Physics will be evaluated by Springer and by members of the Editorial Board. Proposals and other communication should be sent to your Publishing Editors at Springer.

Featuring compact volumes of 50 to 125 pages (approximately 20,000–45,000 words), Briefs are shorter than a conventional book but longer than a journal article. Thus, Briefs serve as timely, concise tools for students, researchers, and professionals.

Typical texts for publication might include:

- A snapshot review of the current state of a hot or emerging field
- A concise introduction to core concepts that students must understand in order to make independent contributions
- An extended research report giving more details and discussion than is possible in a conventional journal article
- A manual describing underlying principles and best practices for an experimental technique
- An essay exploring new ideas within physics, related philosophical issues, or broader topics such as science and society.

Briefs allow authors to present their ideas and readers to absorb them with minimal time investment. Briefs will be published as part of Springer's eBook collection, with millions of users worldwide. In addition, they will be available, just like other books, for individual print and electronic purchase. Briefs are characterized by fast, global electronic dissemination, straightforward publishing agreements, easy-to-use manuscript preparation and formatting guidelines, and expedited production schedules. We aim for publication 8–12 weeks after acceptance.

Ipsita Mandal

# Controlled Description of Non-Fermi Liquids Using Field-Theoretic Methods

Ipsita Mandal
Department of Physics
Shiv Nadar Institution of Eminence
(SNIoE)
Greater Noida, Uttar Pradesh, India

ISSN 2191-5423 ISSN 2191-5431 (electronic)
SpringerBriefs in Physics
ISBN 978-3-032-21617-5 ISBN 978-3-032-21618-2 (eBook)
https://doi.org/10.1007/978-3-032-21618-2



This Springer imprint is published by the registered company Springer Nature Switzerland AG
The registered company address is: Gewerbestrasse 11, 6330 Cham, Switzerland

*I dedicate this book to my mentors and well-wishers, Ashoke Sen, Erich J. Mueller, Emil J. Bergholtz, and Carsten Timm. Their selfless and relentless support has enabled me to achieve this milestone.*

# Foreword

It is with pleasure and a little trepidation that I agreed to write the Foreword to the monograph entitled *Controlled Description of Non-Fermi Liquids Using Field Theoretical Methods* by Dr. Hab. Ipsita Mandal, a brilliant young theorist working now at Shiv Nadar Institution of Eminence in India. It is a pleasure because the book is a pleasure to read, in the choice of the topics, the layout of the book, and the beautiful manner in which the topics have been presented, which is clearly a labour of love by the author. The author is now a condensed matter physicist with basic training in string theory during her thesis work which was done under the supervision of one of the world's leading authorities and celebrated physicist Prof. Ashoke Sen at the Harish-Chandra Institute (HRI), India. This unusual trajectory has led to Dr. Mandal being uniquely qualified to write this tract, which uses methods of field theory to explain condensed matter phenomena.

The book explores non-Fermi liquids (NFLs), a paradigm of strongly coupled phenomena arising in many-body systems. The formulation of the quantum field theory (QFT) techniques is a bit different from what we find in the textbooks of high-energy physics because condensed-matter systems are non-relativistic. A reduced amount of symmetries, for example, compared to those applicable for relativistic fermions, make the calculations harder to execute. The relevant energy-scale flows to be considered are towards the infrared rather than towards the ultraviolet. By studying NFLs arising in myriad systems which, nevertheless, exhibit some robust analogous features, Dr. Mandal emphasises on the overarching universal features in seemingly unrelated systems. Each chapter has an extensive bibliography to aid the readers.

I agreed to write this Foreword because I wear two hats, one as a professional colleague who has familiarity with methods of QFT, and another as a member of the editorial board of the Springer Briefs in Physics on which I have served for a number of years, although this is the first time I am writing one. I normally combine the aforementioned roles by seeking and identifying topics and authors who may be ideally suited to write expository tracts and books and as a colleague and also as a member of the editorial board of a series which serves the important role of bridging the gap between existing literature in the form of well-known textbooks and material that is known to experts in the form of, e.g., review articles or lecture notes. This series plays an important role in bringing out monographs on a short scale while leaving the possibility of a larger and more detailed treatment to the authors for the future. The present book fits neatly in this niche, and it is my belief that the author will also consider such a longer treatment, and for the moment I take this opportunity to congratulate her and the team at Springer including Lisa Scalone, as well as the readers who are sure to embark on a highly enjoyable cruise. I am sure the technical parts will enrich the researcher as well as the student, and the narrative will kindle their desire to learn the subject. It is also my hope that this book will serve as an introductory text to students.

February 2026

Balasubramanian Ananthanarayan
Professor and Former Chairman,
Centre for High Energy Physics
Indian Institute of Science
Bengaluru, India

# Preface

Condensed matter physics is the study of the complex behaviour of a large number of interacting particles such that their collective behaviour gives rise to emergent properties. We will discuss some interesting quantum condensed matter systems wherein intriguing emergent phenomena arise due to strong coupling, showing up as non-Fermi liquids (NFLs). Reviewing the Landau paradigm of Fermi liquid theory, we will understand the distinctive origin of NFLs in contrast with normal metals (or Fermi liquids). We will outline a framework to extract the low-energy physics of such systems in a controlled approximation, using the tool of dimensional regularization. We will demonstrate how this technique can be used to extract the low-energy properties of NFL phases in various strongly correlated systems. In the entire book, we will focus on critical-Fermi-surface states, where there is a well-defined Fermi surface, but no well-defined quasiparticles, because the latter get destroyed as a consequence of strong interactions between the Fermi surface and some emergent massless boson(s). We will focus on the quantum critical points where the dynamics of an order parameter couples with the itinerant fermionic degrees of freedom to cause an NFL behaviour. The intended audience is scientists working in the area of strongly correlated systems. This monograph will also enable researchers, who are new to quantum field theory, to carry out explicitly the basic steps involving dimensional regularization, renormalization group (RG) flows, minimal subtraction scheme, and so on.

Greater Noida, India
February 2026

Ipsita Mandal

**Acknowledgements** The writing of this book was made possible by Balasubramanian Ananthanarayan, who encouraged to take up this endeavour. My knowledge in the area of non-Fermi liquids can mainly be attributed to my association with Sung-Sik Lee, when I joined Perimeter Institute as a Postdoctoral Scholar. I am indebted to him for teaching the nitty-gritty details of the quantum-field-theory (QFT) formulation of such non-relativistic systems. I express my gratitude to my collaborators for various works published on this topic, which include Subir Sachdev, Matthias Punk, Rafael M. Fernandes, Hermann Freire, Kazi Ranjibul Islam, Dimitri Pimenov, and Francesco Piazza. Lastly, I want to express my endless thanks to my Ph.D. supervisor and teacher, Ashoke Sen, from whom I learnt the basics of QFT in the first place.

**Competing Interests** The author has no competing interests to declare that are relevant to the content of this manuscript.

# Contents

# Acronyms

| | |
|---|---|
| CDW | Charge-Density Wave |
| FL | Fermi Liquid |
| FS | Fermi Surface |
| IR | Infrared |
| MS | Minimal Subtraction Scheme |
| NFL | Non-Fermi Liquid |
| QCP | Quantum Critical Point |
| QFT | Quantum Field Theory |
| RG | Renormalization Group |
| UV | Ultraviolet |

# Chapter 1
# Introduction

Non-Fermi liquids (NFLs) are metallic states that encapsulate, in striking fashion, the consequences of strong interactions in condensed matter systems [1–14], and resist analysis by conventional theoretical methods. Chief among these is Landau's Fermi liquid (FL) theory, which provides an excellent description of normal metals but breaks down entirely for NFLs. The reason is fundamental: Landau's framework rests on the existence of long-lived quasiparticles, and when interactions are strong enough to destroy these quasiparticles, the entire edifice collapses. NFLs therefore exhibit thermodynamic and transport properties [15–20] that are qualitatively distinct from those of ordinary metals. A telling example is the DC resistivity $\rho$: in a conventional FL, electron-electron interactions give rise to the characteristic $\rho \sim T^2$ dependence at low temperatures, whereas in an NFL one finds $\rho \sim T^{\alpha}$ with $\alpha \neq 2$, the most common cases being $\alpha = 1$ and $\alpha = 4/3$. Further distinctions manifest in properties such as the optical conductivity [15, 16] and an enhanced propensity toward superconducting instabilities [16, 21, 22]. The principal physical scenarios in which such strange-metallic behavior emerges include: (1) finite-density fermions interacting with order-parameter bosons that become massless at a quantum critical point (QCP) [1, 2, 4–7, 11–13]; (2) finite-density fermions coupled to transverse gauge fields [3, 8, 21, 23] that remain massless; and (3) Fermi points at band-touching semimetals subject to unscreened Coulomb interactions, probed at zero chemical potential [17–20, 24–29]. This last class involves NFL behaviour at isolated Fermi points (encoding points where the density-of-states goes to zero) and is therefore free from the complications associated with the presence of an extended Fermi surface (FS) accompanying finite-density systems.

Extensive efforts have been devoted to accessing NFL states through controlled approximations, employing a range of distinct and complementary approaches, which we now survey in turn:

1. The approach of Hertz and Millis [30, 31] proceeds by performing a renormalization group (RG) analysis on an effective bosonic action for the order parameter, obtained by integrating out the fermions. This procedure is not well-justified,

 
I. Mandal, *Controlled Description Of Non-Fermi Liquids Using Field-Theoretic Methods*, SpringerBriefs in Physics, https://doi.org/10.1007/978-3-032-21618-2_1

however, since one set of gapless degrees of freedom—the fermionic fields—is eliminated at the outset, and both fermions and bosons must instead be treated on an equal footing.

2. Setting aside the Hertz-Millis formulation, an early and natural attempt involved introducing a large number $N$ of fermion flavors [32–34] as a mathematical device, motivated by the intuition that additional flavors should not qualitatively alter the physics beyond enlarging the flavor-symmetry group. The infinite-flavor limit was then to serve as the starting point for a controlled $1/N$ expansion, with the hope that setting $N$ to its physical value at the end would yield a reliable approximation. It was subsequently established, however, that the $N \to \infty$ limit is not described by a mean-field theory for finite-density fermions with a well-defined FS [1, 35], contrary to the original assumption. The culprit is the proliferation of planar diagrams [35], which renders even the large-$N$ theory strongly interacting: solving it requires a nontrivial resummation of infinitely many Feynman diagrams. These serious shortcomings prompted the search for alternative approaches.
3. The insights above also call into question physical pictures based on the random-phase approximation (RPA), since RPA amounts to nothing more than the leading-order term in the large-$N$ expansion. One alternative scheme that circumvents these difficulties involves deforming the bare dispersion of the quantum fields. In Refs. [36, 37], this was accomplished by replacing the bosonic kinetic term $\boldsymbol{k}^2|\phi(k)|^2$ with $\boldsymbol{k}^{1+\epsilon}|\phi(k)|^2$. For $\epsilon \in (0, 1)$, the density of states of the order-parameter boson is suppressed at low energies, reducing quantum fluctuations and making $\epsilon$ a viable perturbative parameter. This scheme retains a finite fermionic density of states and preserves all microscopic symmetries. Its drawback, however, is that the nonanalytic momentum dependence of the bosonic kinetic term corresponds to a nonlocal hopping in real space, which prevents the collective mode from acquiring an anomalous dimension, since short-distance quantum fluctuations cannot renormalize a nonlocal hopping term.
4. The approach that achieves genuine mathematical control over the perturbative expansion is dimensional regularization [3–5]. In our non-relativistic setting, one can tune independently the number of dimensions perpendicular to the FS (denoted as $d_{co}$) and the dimension of the FS itself (denoted as $d_F$). Increasing $d_{co}$ amounts to continuously raising the number of spacetime dimensions in which the FS is embedded, which suppresses quantum fluctuations by reducing the density of states in higher dimensions.

Throughout this book, we adopt the dimensional regularization framework to analyse the systems under consideration, in the spirit of Refs. [4, 5]. A key virtue of this approach is that it preserves locality in real space and, for $d_F = 1$, an additional locality emerges in the momentum space as well. The next chapter lays out the most important building blocks for constructing the effective action that serves as the starting point for applying dimensional regularization, with the Ising-nematic quantum critical point as the guiding example. The NFL systems treated in the remaining chapters draw on minor variations of this framework, with the essential structure established in Chap. 2 remaining intact throughout. The variation mainly lies in the formulation of the two-component spinors, which lead to differing forms of the one-loop bosonic self-energy. Despite the different forms, all of them feature a term representing Landau-damping (analogous to an overdamped harmonic-oscillator mode) arising from one-loop corrections. The overarching consequence of this is the drastic modification of the dependence of the fermionic self-energy on the Matsubara-frequency ($k_0$) from the $\sim k_0^2$ behaviour of the FLs—the resulting one-loop fermionic self-energy shows a universal scaling form, viz. $\sim \text{sgn}(k_0)|k_0|^{2/3}$.

## References

1. M.A. Metlitski, S. Sachdev, Phys. Rev. B **82**, 075127 (2010). https://doi.org/10.1103/PhysRevB.82.075127
2. M.A. Metlitski, S. Sachdev, Phys. Rev. B **82**, 075128 (2010). https://doi.org/10.1103/PhysRevB.82.075128
3. S. Chakravarty, R.E. Norton, O.F. Syljuåsen, Phys. Rev. Lett. **74**, 1423 (1995). https://doi.org/10.1103/PhysRevLett.74.1423. https://link.aps.org/doi/10.1103/PhysRevLett.74.1423
4. D. Dalidovich, S.S. Lee, Phys. Rev. B **88**, 245106 (2013). https://doi.org/10.1103/PhysRevB.88.245106
5. I. Mandal, S.S. Lee, Phys. Rev. B **92**, 035141 (2015). https://doi.org/10.1103/PhysRevB.92.035141
6. I. Mandal, Eur. Phys. J. B **89**(12), 278 (2016). https://doi.org/10.1140/epjb/e2016-70509-4
7. D. Pimenov, I. Mandal, F. Piazza, M. Punk, Phys. Rev. B **98**, 024510 (2018). https://doi.org/10.1103/PhysRevB.98.024510
8. I. Mandal, Phys. Rev. Res. **2**, 043277 (2020). https://doi.org/10.1103/PhysRevResearch.2.043277. https://link.aps.org/doi/10.1103/PhysRevResearch.2.043277
9. T. Holder, W. Metzner, Phys. Rev. B **90**, 161106 (2014). https://doi.org/10.1103/PhysRevB.90.161106. https://link.aps.org/doi/10.1103/PhysRevB.90.161106
10. J. Sýkora, T. Holder, W. Metzner, Phys. Rev. B **97**, 155159 (2018). https://doi.org/10.1103/PhysRevB.97.155159. https://link.aps.org/doi/10.1103/PhysRevB.97.155159
11. I. Mandal, Nucl. Phys. B **1005**, 116586 (2024). https://doi.org/10.1016/j.nuclphysb.2024.116586
12. I. Mandal, Ann. Phys. **474**, 169925 (2025). https://doi.org/10.1016/j.aop.2025.169925
13. I. Mandal, R.M. Fernandes, Phys. Rev. B **107**, 125142 (2023). https://doi.org/10.1103/PhysRevB.107.125142. https://link.aps.org/doi/10.1103/PhysRevB.107.125142
14. I. Mandal, Ann. Phys. **474**, 169925 (2025). https://doi.org/10.1016/j.aop.2025.169925. https://www.sciencedirect.com/science/article/pii/S0003491625000065
15. A. Eberlein, I. Mandal, S. Sachdev, Phys. Rev. B **94**, 045133 (2016). https://doi.org/10.1103/PhysRevB.94.045133. http://link.aps.org/doi/10.1103/PhysRevB.94.045133
16. I. Mandal, Ann. Phys. **376**, 89 (2017). https://doi.org/10.1016/j.aop.2016.11.009. http://www.sciencedirect.com/science/article/pii/S0003491616302585
17. I. Mandal, H. Freire, Phys. Rev. B **103**, 195116 (2021). https://doi.org/10.1103/PhysRevB.103.195116. https://link.aps.org/doi/10.1103/PhysRevB.103.195116
18. H. Freire, I. Mandal, Phys. Lett. A **407**, 127470 (2021). https://doi.org/10.1016/j.physleta.2021.127470. https://www.sciencedirect.com/science/article/pii/S0375960121003340
19. I. Mandal, H. Freire, J. Phys.: Condens. Matter **34**(27), 275604 (2022). https://doi.org/10.1088/1361-648x/ac6785. https://doi.org/10.1088/1361-648x/ac6785
20. I. Mandal, H. Freire, J. Phys.: Condens. Matter **36**(44), 443002 (2024). https://doi.org/10.1088/1361-648X/ad665e. https://dx.doi.org/10.1088/1361-648X/ad665e
21. S.B. Chung, I. Mandal, S. Raghu, S. Chakravarty, Phys. Rev. B **88**, 045127 (2013). https://doi.org/10.1103/PhysRevB.88.045127. http://link.aps.org/doi/10.1103/PhysRevB.88.045127

22. I. Mandal, Phys. Rev. B **94**, 115138 (2016). https://doi.org/10.1103/PhysRevB.94.115138
23. Z. Wang, I. Mandal, S.B. Chung, S. Chakravarty, Ann. Phys. **351**, 727 (2014). https://doi.org/10.1016/j.aop.2014.09.021. http://www.sciencedirect.com/science/article/pii/S0003491614002814
24. A.A. Abrikosov, Sov. Phys.-JETP **39**, 709 (1974)
25. E.G. Moon, C. Xu, Y.B. Kim, L. Balents, Phys. Rev. Lett. **111**, 206401 (2013). https://doi.org/10.1103/PhysRevLett.111.206401. https://link.aps.org/doi/10.1103/PhysRevLett.111.206401
26. R.M. Nandkishore, S.A. Parameswaran, Phys. Rev. B **95**, 205106 (2017). https://doi.org/10.1103/PhysRevB.95.205106. https://link.aps.org/doi/10.1103/PhysRevB.95.205106
27. I. Mandal, R.M. Nandkishore, Phys. Rev. B **97**(12), 125121 (2018). https://doi.org/10.1103/PhysRevB.97.125121
28. B. Roy, M.P. Kennett, K. Yang, V. Juričić, Phys. Rev. Lett. **121**, 157602 (2018) https://doi.org/10.1103/PhysRevLett.121.157602. https://link.aps.org/doi/10.1103/PhysRevLett.121.157602
29. I. Mandal, Phys. Lett. A **418**, 127707 (2021). https://doi.org/10.1016/j.physleta.2021.127707. https://www.sciencedirect.com/science/article/pii/S0375960121005715
30. J.A. Hertz, Phys. Rev. B **14**, 1165 (1976). https://doi.org/10.1103/PhysRevB.14.1165. https://link.aps.org/doi/10.1103/PhysRevB.14.1165
31. A.J. Millis, Phys. Rev. B **48**, 7183 (1993). https://doi.org/10.1103/PhysRevB.48.7183. https://link.aps.org/doi/10.1103/PhysRevB.48.7183
32. B.L. Altshuler, L.B. Ioffe, A.J. Millis, Phys. Rev. B **50**, 14048 (1994). https://doi.org/10.1103/PhysRevB.50.14048. http://link.aps.org/doi/10.1103/PhysRevB.50.14048
33. J. Polchinski, Nucl. Phys. B **422**, 617 (1994). https://doi.org/10.1016/0550-3213(94)90449-9
34. E.A. Kim, M.J. Lawler, P. Oreto, S. Sachdev, E. Fradkin, S.A. Kivelson, Phys. Rev. B **77**(18), 184514 (2008). https://doi.org/10.1103/PhysRevB.77.184514
35. S.S. Lee, Phys. Rev. B **80**, 165102 (2009). https://doi.org/10.1103/PhysRevB.80.165102
36. C. Nayak, F. Wilczek, Nucl. Phys. B **430**, 534 (1994). https://doi.org/10.1016/0550-3213(94)90158-9
37. D.F. Mross, J. McGreevy, H. Liu, T. Senthil, Phys. Rev. B **82**, 045121 (2010). https://doi.org/10.1103/PhysRevB.82.045121

# Chapter 2
# Ising-Nematic Quantum Critical Point

## 2.1 Introduction

In this chapter, we focus our attention to a particularly illuminating example: an Ising-nematic order parameter that couples with the quadrupolar distortion of a Fermi surface (FS). At a quantum phase transition, this coupling drives the spontaneous breaking of a fourfold-symmetric FS down to a twofold-symmetric one, and the system crosses through what is known as a non-Fermi liquid (NFL) phase. Our goal in what follows is to show how a controlled quantum field-theoretic analysis—employing the powerful technique of dimensional regularization—can be used to determine the low-energy scaling behavior of such NFLs embedded in a generic spatial dimension symbolised by $d$.

For a globally convex FS in momentum space, no single point on the surface is privileged over any other, and the geometry is naturally characterized by two quantities: the dimension $m$ of the FS itself, and its co-dimension $\tilde{d} = d - m$. The virtue of dimensional regularization is that $m$ and $\tilde{d}$ can be varied independently. Roughly speaking, $d$ controls how strongly quantum fluctuations are felt, while $m$ governs how many gapless modes are available. In real physical systems, $d$ and $m$ are of course positive integers (with $d \geq 2$ and $m \geq 1$), but we shall treat them as continuously tunable parameters—a mathematical artifice that allows us to smoothly approach the physical dimension $d_p$ where the relevant degrees of freedom are strongly coupled. When $d_p$ lies below the upper critical dimension $d_c$, the system flows at low energies toward an interacting NFL; when $d_p > d_c$, a conventional Fermi-liquid (FL) description remains adequate.

Among the NFLs that arise when $d_p < d_c$, there is a qualitative distinction between the cases $m = 1$ and $m > 1$, and the origin of this distinction is worth dwelling on. When $m = 1$, an emergent locality appears in momentum space: observables such as Green's functions, which are local in momentum, can be extracted from small patches of the FS without any reference to its global structure. This locality is what makes controlled NFL descriptions possible in the so-called patch formulation. When

 
I. Mandal, *Controlled Description Of Non-Fermi Liquids Using Field-Theoretic Methods*, SpringerBriefs in Physics, https://doi.org/10.1007/978-3-032-21618-2_2

$m > 1$, however, this locality is lost. The size of the FS, set by $k_F$, enters the bosonic propagator through the Landau-damping term and qualitatively modifies the scaling behavior. This is a manifestation of UV/IR mixing—a situation in which low-energy physics is influenced by gapless modes spread across the entire FS, and therefore cannot be captured purely by renormalizing local properties near any given point. In this sense, $k_F$ becomes a "naked scale": it resists being absorbed into the local effective description and leaves a permanent imprint on the low-energy theory.

To appreciate why UV/IR mixing is so unusual, it helps to contrast it with the familiar situation in renormalizable relativistic quantum field theories. In QED, for instance, any observable at a momentum $|\boldsymbol{k}_1| \ll \Lambda$ can be expressed entirely in terms of the renormalized mass and charge measured at some other low-energy scale $|\boldsymbol{k}_2| \ll \Lambda$, up to corrections suppressed by powers of $|\boldsymbol{k}_1|/\Lambda$. Long-distance and short-distance physics decouple cleanly. This clean decoupling breaks down in the presence of a FS, because $k_F$ introduces an additional energy scale that satisfies $k_F \gg \Lambda$, and when $m > 1$, low-energy observables near the FS can no longer be expressed solely in terms of effective couplings defined locally in momentum space.

One might hope to evade this complication by working with a critical boson possessing a large number of flavours, or a velocity much greater than the Fermi velocity, since either limit pushes the effects of $k_F$ to very low energies. Yet UV/IR mixing is ultimately unavoidable: as long as the number of flavours and the velocity ratio remain finite, it reasserts itself at sufficiently low energies, either through Landau-damping or through the onset of a superconducting instability. With this motivation in hand, the remainder of the chapter is devoted to a controlled field-theoretic treatment of the NFL arising at the Ising-nematic quantum critical point (QCP) for an $m$-dimensional FS embedded in $d$ spatial dimensions—one that carefully tracks how interactions and UV/IR mixing together conspire to shape the low-energy scaling behavior.

## 2.2 Model

We consider an $m$-dimensional FS coupled to a critical boson whose momentum is centered at $\mathbf{Q} = 0$, embedded in $d_p = m + 1$ spatial dimensions. One natural way to characterize the resulting NFL is through the scaling behavior of the fermionic and bosonic Green's functions. For this purpose, it is convenient to anchor the analysis at a particular point on the Fermi surface—call it $K^*$—at which the fermionic Green's function is defined. At low energies, fermions scatter predominantly along the tangential directions of the Fermi surface while interacting with the critical boson. In the presence of inversion symmetry, fermions in the vicinity of $K^*$ couple most strongly to fermions near the antipodal point $-K^*$, whose tangent space coincides with that of $K^*$. With this observation in mind, we divide the closed Fermi surface into two halves centered at $K^*$ and $-K^*$, and introduce separate fermionic fields $\psi_{+,\lambda}$ and $\psi_{-,\lambda}$ to represent the degrees of freedom on each half, as illustrated in Fig. 2.1. In this

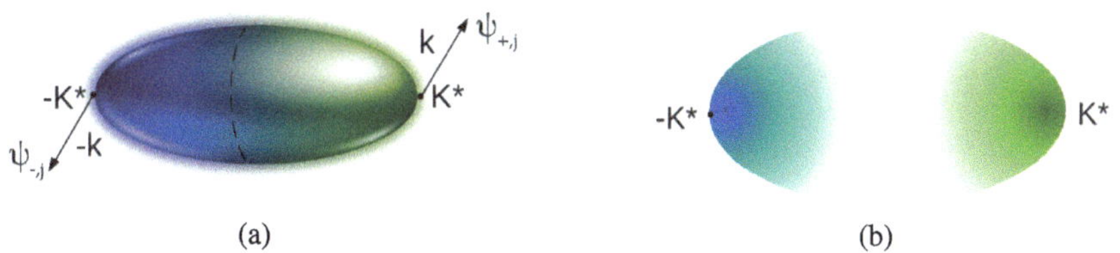


**Fig. 2.1** **a** A compact Fermi surface can be divided into two halves, centered at $\pm K^*$. In the vicinity of each centre, a separate fermionic field is introduced. **b** The compact Fermi surface is approximated by two sheets of non-compact Fermi surfaces, with a momentum regularization that suppresses modes lying

coordinate system, the action written in Matsubara-frequency space at temperature $T = 0$ takes the form of

$$
\begin{aligned}
S = &\sum_{\lambda,\, s=\pm} \int dk\, \psi^\dagger_{s,\lambda}(k)\Big[i\,k_0 + s\,k_1 + \mathbf{L}^2_{(k)} + H\big(\mathbf{L}^2_{(k)}\big)\Big]\psi_{s,\lambda}(k) \\
&+ \frac{1}{2}\int dk\,\big[k_0^2 + c_\perp\, k_1^2 + c_\parallel \mathbf{L}^2_{(k)}\big]\,\phi(-k)\,\phi(k) \\
&+ \frac{1}{\sqrt{N}} \sum_{\lambda,\, s=\pm} \int dk\, dq\; e\,\phi(q)\; \psi^\dagger_{s,\lambda}(k+q)\,\psi_{s,\lambda}(k).
\end{aligned} \tag{2.1}
$$

Here, $k \equiv \{k_0,\, \boldsymbol{k}\}$ denotes the $(d+1)$-dimensional energy-momentum vector, where $\boldsymbol{k}$ comprises the components $\{k_j\}$ with $1 \le j \le d$, and $dk \equiv d^{d+1}k/(2\pi)^{d+1}$ is the corresponding integration measure. The fields $\psi_{+,\lambda}(k)$ and $\psi_{-,\lambda}(k)$ represent fermionic degrees of freedom carrying an additional flavor index $\lambda \in \{1, 2, \ldots, N\}$, frequency $k_0$, and momenta $\boldsymbol{K}^* + \boldsymbol{k}$ and $-\boldsymbol{K}^* + \boldsymbol{k}$, respectively. The component $k_1$ denotes the momentum direction perpendicular to the Fermi surface at $\pm K^*$, while $\mathbf{L}_{(k)} \equiv (k_2, k_3, \ldots, k_d)$ collects the components running parallel to it. The momentum is rescaled so that both the magnitude of the Fermi velocity and the quadratic curvature of the Fermi surface at $\pm K^*$ are normalized to unity. Since the FS is locally parabolic in this neighborhood, it is natural to assign scaling dimensions of 1 and $1/2$ to $k_1$ and $\mathbf{L}_{(k)}$, respectively. The function

$$
H\big(\mathbf{L}^2_{(k)}\big) = \sum_{n=3}^{\infty} \sum_{i_1,\ldots,i_n=2}^{d} c_{i_1,\ldots,i_n} \frac{k_{i_1}\cdots k_{i_n}}{k_F^{\frac{n-2}{2}}} \tag{2.2}
$$

collects all cubic and higher-order terms in $\mathbf{L}_{(k)}$, where $k_F$ is a scale of scaling dimension 1. The integration range of $\mathbf{L}_{(k)}$ in $\int dk$ is set by the extent of the Fermi surface, which is of order $\sqrt{k_F}$ in this coordinate system. Finally, the bosonic field $\phi$ is a real scalar that couples to the fermions through a Yukawa-like interaction. This

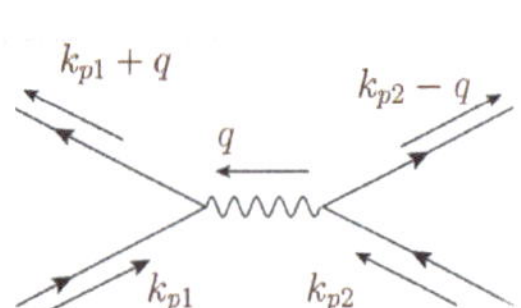


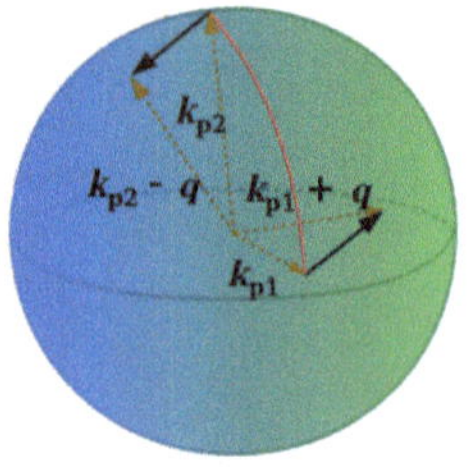


**Fig. 2.2** For a FS with dimension $m > 1$, any two points on the FS have a commom tangent space which is $(m-1)$-dimensional

coupling is obtained by expanding the quadrupolar distortion of the Fermi surface around $\pm K^*$ and retaining only the leading-order, momentum-independent term [1, 2].

Although the action bears a resemblance to the patch theories that have been used to describe NFLs for $m = 1$ [3], the situation for $m > 1$ is qualitatively richer: any two points on the Fermi surface share at least $(m-1)$ common tangent vectors. Figure 2.2 illustrates this for the concrete case of a spherical Fermi surface embedded in three-dimensional momentum space. Consider two fermions carrying momenta $\boldsymbol{k}_{p1}$ and $\boldsymbol{k}_{p2}$, which scatter to $\boldsymbol{k}_{p1}+\boldsymbol{q}$ and $\boldsymbol{k}_{p2}-\boldsymbol{q}$ by exchanging a boson with small momentum $\boldsymbol{q}$. When $m > 1$, the transferred momentum $\boldsymbol{q}$ can be tangential to the Fermi surface simultaneously at both $\boldsymbol{k}_{p1}$ and $\boldsymbol{k}_{p2}$, which means that both fermions remain near the Fermi surface before and after the scattering event. As a result, any two fermionic fields on the Fermi surface stay strongly coupled in the low-energy limit, even though processes involving large momentum transfers are suppressed. This stands in sharp contrast to the $m = 1$ case, where such low-energy scattering is absent except between antipodal points. Because these interactions are spread nonlocally across the entire FS in the momentum space, the theory with $m > 1$ is ill-defined in the $k_F \to \infty$ limit, unlike the $m = 1$ case. Put differently, the low-energy (IR) observables, such as the fermionic and bosonic Green's functions near $k = 0$, cannot be determined until the global properties of the Fermi surface—its size and shape at large momenta—are specified. This is the essence of UV/IR mixing, which in the Fermi liquid context is encoded in the Landau parameters, themselves nonlocal objects in momentum space.

The scale $k_F$ appearing in $H(\mathbf{L}^2_{(k)})$ provides a large-momentum cutoff along the directions parallel to the FS. Although these higher-order terms are irrelevant by naive power counting, they play an essential role in encoding the compactness of the Fermi surface and cannot simply be discarded. In principle, retaining this information requires an infinite set of independent parameters $\{c_{i_1,\dots,i_n}\}$, each encoding a different aspect of the Fermi surface shape away from $K^*$. Rather than working with this full complexity, we instead adopt a simplified ultraviolet regularization that captures the essential physics of these higher-order terms while remaining tractable analytically. Concretely, we replace the kinetic term with a regularized version,

$$\sum_{s,\lambda} \int dk\, \psi^\dagger_{s,\lambda}(k) \Big[ i\, k_0 + s\, k_{d-m} + \mathbf{L}^2_{(k)} \Big] \psi_{s,\lambda}(k)\, \exp\left\{ \frac{\mathbf{L}^2_{(k)}}{k_F} \right\}. \tag{2.3}$$

Here the dispersion remains parabolic, but the exponential factor renders the FS effectively finite by suppressing the kinematics of fermions with $|\mathbf{L}_{(k)}| > \sqrt{k_F}$, as illustrated in Fig. 2.1.

In order to control the Yukawa-like coupling between the fermions and bosons, and the strength of the UV/IR mixing, independently of one another, we tune both $d$ [4–6] and $m$ [3, 7, 8]. To maintain the analyticity of the theory in momentum space—which is equivalent to locality in real space—for generic values of $m$, we introduce a two-component spinor [3, 7],

$$\Psi_j^T(k) = \big[ \psi_{+,\lambda}(k) \quad \psi^\dagger_{-,\lambda}(-k) \big], \tag{2.4}$$

and use it to formulate an action embedded in a $d$-dimensional momentum space, which takes the form of

$$\begin{aligned} S = &\sum_\lambda \int dk\, \bar{\Psi}_j(k)\, i \Big[ \mathbf{\Gamma} \cdot \mathbf{K} + \gamma_{d-m}\, \delta_k \Big] \Psi_j(k)\, e^{\frac{\mathbf{L}^2_{(k)}}{\mu k_F}} + \frac{1}{2} \int dk\, \mathbf{L}^2_{(k)}\, \phi(-k)\, \phi(k) \\ &+ \frac{i\, e\, \mu^{x_e/2}}{\sqrt{N}} \sum_\lambda \int dk dq\, \phi(q)\, \bar{\Psi}_j(k+q)\, \gamma_{d-m} \Psi_j(k). \end{aligned} \tag{2.5}$$

Here, $\mathbf{K} \equiv \{k_0, k_1, \ldots, k_{d-m-1}\}$ collects the frequency and the first $(d-m-1)$ components of the $d$-dimensional momentum vector, while $\mathbf{L}_{(k)} \equiv \{k_{d-m+1}, \ldots, k_d\}$ and $\delta_k = k_{d-m} + \mathbf{L}^2_{(k)}$. Within the $d$-dimensional momentum space, the components $\{k_1, \ldots, k_{d-m}\}$ and $\mathbf{L}_{(k)}$ represent the $(d-m)$ and $m$ directions perpendicular and parallel to the FS, respectively. The vector $\mathbf{\Gamma} \equiv \{\gamma_0, \gamma_1, \ldots, \gamma_{d-m-1}\}$ collects the gamma-matrices associated with $\mathbf{K}$. Since we are interested in co-dimensions satisfying $1 \leq d - m \leq 2$, it suffices to work with $2 \times 2$ gamma-matrices, for which we take $\gamma_0 = \sigma_y$, $\gamma_{d-m} = \sigma_x$, and define $\bar{\Psi} \equiv \Psi^\dagger \gamma_0$.

The leading terms in the non-interacting part of Eq. (2.5) are invariant under the following rescaling transformations, expressed in terms of a mass scale $\mu$:

$$\begin{aligned} &\mathbf{K} = \mu\, \mathbf{K}', \quad k_{d-m} = \mu\, k'_{d-m}, \quad \mathbf{L}_{(k)} = \sqrt{\mu}\, \mathbf{L}'_{(k)}, \\ &\Psi_j(k) = \mu^{-\frac{2d+4-m}{4}}\, \Psi'_j(k'), \quad \phi(k) = \mu^{-\frac{2d+4-m}{4}}\, \phi'(k'). \end{aligned}$$

In the quadratic action of the boson, only the term $\mathbf{L}^2_{(k)}\, \phi^*(k)\phi(k)$ is retained, because $|\mathbf{K}|^2 + k^2_{d-m}$ is irrelevant under the scaling in which each of $\{k_0, k_1, \ldots, k_{d-m}\}$ carries dimension 1 and each of $\{k_{d-m+1}, \ldots, k_d\}$ carries dimension $1/2$. Put differently, the bosonic dynamics is so strongly dressed by particle-hole excitations that portions of the bare kinetic term become unimportant in the infrared. While the $(m+1)$-dimensional rotational symmetry of the bosonic action would naively place all of $\{k_{d-m}, \ldots, k_d\}$ on equal footing, the coupling to the fermions singles out $k_{d-m}$: the bosons that contribute most to fermionic scattering around $\pm K^*$ carry momenta satisfying $|\mathbf{L}_{(k)}| \gg k_{d-m}$, and so the dependence on $k_{d-m}$ in the bosonic kinetic term can be safely neglected when describing the fermionic dynamics in that region. Finally, $c_{\parallel}$ has been absorbed into a redefinition of the field.

The scaling dimension of the Yukawa-like coupling constant $e_0 = e\,\mu^{x_e/2}$ is $x_e/2$, where $x_e = 2 + m/2 - d$. By writing $e_0 = e\,\mu^{x_e/2}$, where $\mu$ is a mass scale, the coupling $e$ is rendered conveniently dimensionless. We further define $\tilde{k}_F = k_F/\mu$ as the dimensionless counterpart of $k_F$. The spinor carries an energy dispersion with two bands, $E_k = \pm\sqrt{\sum_{j=1}^{d-m-1} k_j^2 + \delta_k^2}$, which gives rise to an $m$-dimensional FS embedded in a $d$-dimensional momentum space, defined by the $d-m$ conditions $k_i = 0$ for $i \in \{1, \ldots, d-m-1\}$ and $k_{d-m} = -\mathbf{L}_{(k)}^2$.

Apart from $k_F$, which sets the size of the FS, the theory implicitly carries a UV cut-off $\Lambda$ for $\mathbf{K}$ and $k_{d-m}$. It is natural to identify $\Lambda = \mu$, which sets the largest energy—or equivalently, the largest momentum perpendicular to the FS—that fermions may carry. If $k_e$ denotes the typical energy scale at which one wishes to probe the system, the regime of interest is $k_e \ll \Lambda \ll k_F$. We study the RG flows of the two dimensionless parameters $e$ and $\tilde{k}_F$, generated by varying $\Lambda$ while requiring that low-energy observables remain independent of it. This is equivalent to the Wilsonian procedure of coarse-graining, in which high-energy modes away from the FS are integrated out at each step. Since the zero-energy modes are never integrated out, the ratio $k_F/\Lambda$ grows monotonically throughout the coarse-graining procedure. We therefore treat $k_F$ as a dimensionful coupling constant that flows to infinity in the low-energy limit—a fact that captures, physically, the divergence of the FS size measured in units of the thickness of the thin shell surrounding it, as illustrated in Fig. 2.3a.

## 2.3 Dimensional Regularization

For a given $m$, we tune $d$ towards the critical dimension, $d_c$, which defines the value of $d$ at which the one-loop quantum corrections diverge logarithmically in $\Lambda$, where $\Lambda \ll k_F$. Clearly, $x_e$ vanishes when $\tilde{d}_c(m) = 2 + m/2$ and, in general, it will turn out that $\tilde{d}_c \neq d_c$ for $1 < m < 2$. This mismatch is a sign that $k_F$ enters the low-energy physics in a way that is singular in the large $k_F$ limit, resulting in UV/IR mixing. In order to identify $d_c$, we consider the one-loop quantum corrections. Since the bare bosonic propagator is independent of $\{k_0, \ldots, k_{d-m}\}$, the loop integrations involving it are ill-defined, unless one resums a series of diagrams that provides a non-trivial dispersion along those directions. This amounts to rearranging the perturbative expansion such that the one-loop bosonic self-energy (cf. Fig. 2.4a), $\Pi_1(k) = -(i\,e)^2\,\mu^{x_e} \int dq\, \mathrm{Tr}\left[\gamma_{d-m}\, G_0(k+q)\, \gamma_{d-m}\, G_0(q)\right]$, is included at the 'zero'-th order. Our unusual ordering of including Feynman diagrams, which is not organized by the number of loops, is forced upon us by the dynamical structure

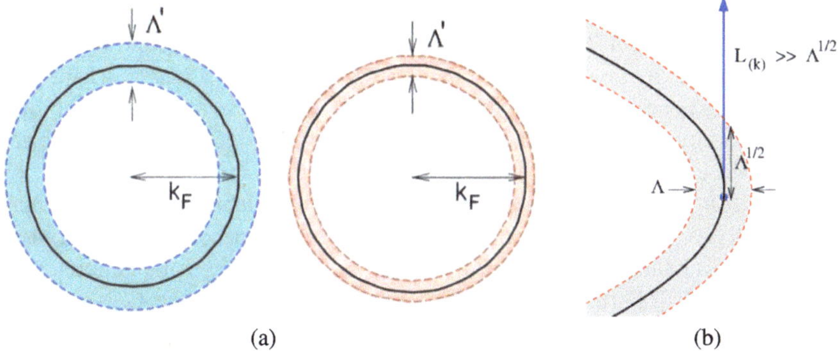


**Fig. 2.3** **a** As the high-energy modes away from the FS are integrated out, the ratio $k_F/\Lambda$ grows, where $k_F$ denotes the size of the FS and $\Lambda$ is the energy cutoff perpendicular to it. For $m > 1$, the Green's functions develop singularities in the $k_F/\Lambda \to \infty$ limit, and it is precisely this behavior that gives rise to the UV/IR mixing of energy scales. **b** A two-dimensional slice of an $m$-dimensional FS. The typical momentum carried by a boson is proportional to $\tilde{\alpha}^{1/3}\,\Lambda^{\frac{d-m}{3}} \sim \tilde{e}^{\frac{1}{m+1}}\,(k_F/\Lambda)^{\frac{m-1}{2(m+1)}}\,\Lambda^{1/2}$. For $m > 1$, this momentum greatly exceeds $\Lambda^{1/2}$ in the low-energy limit. Consequently, the momentum transferred by a boson is large enough to carry a fermion near the FS well outside the thin shell of width $\sim \Lambda$, leading to a suppression of virtual particle-hole excitations by powers of $\Lambda/k_F$ for $m > 1$

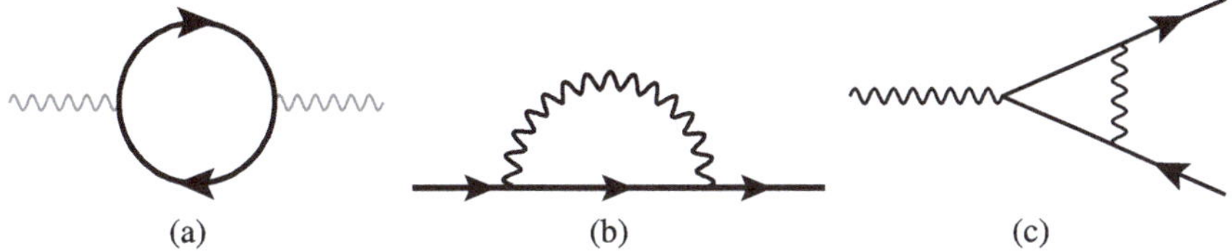


**Fig. 2.4** The one-loop diagrams for the **a** bosonic self-energy, **b** fermionic self-energy, and **c** vertex correction. In (**a**), the gray wiggly curve denotes the bare bosonic propagator. The solid arrows represent bare fermionic propagators, while the wiggly curves in (**b**) and (**c**) denote dressed bosonic propagators that incorporate the one-loop self-energy shown in (**a**)

of the theory. Nevertheless, we adopt a systematic procedure which guarantees that every Feynman diagram is included once and only once order by order in $\epsilon \equiv d_c - d_p$ (Fig. 2.4).

The dressed boson propagator is the one which includes the one-loop self-energy and is expressed as

$$D_1(k) = \left[\mathbf{L}_{(k)}^2 - \Pi_1(k)\right]^{-1}, \quad \Pi_1(k) = -\,\beta_d\, e^2\, \mu^{x_e}\, \frac{(\mu\, \tilde{k}_F)^{\frac{m-1}{2}}\, |\mathbf{K}|^{d-m}}{|\mathbf{L}_{(k)}|},$$

$$\beta_d = \frac{\Gamma^2(\frac{d-m+1}{2})}{2^{\frac{2d+m-1}{2}}\, \pi^{\frac{d-1}{2}}\, |\cos\{\frac{\pi(d-m+1)}{2}\}|\, \Gamma(\frac{d-m}{2})\Gamma(d-m+1)}, \tag{2.6}$$

to the leading order in $k/k_F$, for $|\mathbf{K}|^2/|\mathbf{L}_{(k)}|^2$, $\delta_k^2/|\mathbf{L}_{(k)}|^2 \ll k_F$. The constant $\beta_d$ is finite for $2 \leq d < 3$. Since $D_1(k)$ depends on $e$, the higher-loop diagrams are not accompanied by powers of $e^2$, but rather by powers of $\tilde{e} = e^{fr}$, where $fr$ is a fractional exponent [3]. The nonanalyticity of the exponent in the definition of the effective coupling signals that part of the quantum effects of the bosonic self-energy have been incorporated nonperturbatively through a resummation of the loop diagrams. It is worth emphasizing that $-\Pi_1(k)$ is the celebrated *Landau-damping* term, which gives rise to the characteristic $\text{sgn}(k_0)\,|k_0|^{2/3}$ dependence of the fermionic self-energy—a hallmark of NFL behaviour across a wide range of quantum critical systems [3, 9–15]. Furthermore, $D_1(k)$ diverges for $m > 1$ in the $k_F \to \infty$ limit. This reflects the fact that Landau-damping grows stronger as the FS becomes larger when $m > 1$, since a boson can then decay into particle-hole excitations spread across the entire FS. This stands in sharp contrast to the $m = 1$ case, where a low-energy boson with a given momentum can only decay into particle-hole excitations near isolated patches whose tangents are (anti)parallel to its momentum. For $m = 1$ and $m = 2$, $k_F$ drops out of Eq. (2.6), signaling the absence of UV/IR mixing. We note that Eq. (2.6) is valid only when at least one direction remains tangential to the FS, and should therefore not be extended to $m < 1$, where conventional QFT methods apply without difficulty. In what follows, we restrict our attention to $m > 1$.

The apparent breakdown of the spatial rotational symmetry in the space spanned by the momentum coordinates, $\{k_{d-m}, \ldots, k_d\}$, [cf. Eq. (2.6)] is an artifact of the fact that the expression is valid only for bosons whose momentum is nearly tangential to the FS at $\pm K^*$. For a boson propagator with a generic momentum, $|\mathbf{L}_{(k)}|$ in Eq. (2.6) should be replaced by $\sqrt{k_{d-m}^2 + \cdots + k_d^2}$, as required by the $(m+1)$-dimensional rotational symmetry. This is because, for any given boson momentum $k$, one can always find a point on the FS where $k$ is the local tangent. Choosing a coordinate system in which $k_{d-m} = 0$, the bosonic self-energy takes exactly the same form as in Eq. (2.6), and since this holds for any $k$, a generic boson propagator must be independent of the direction within the subspace spanned by $\{k_{d-m}, \ldots, k_d\}$. In what follows, we work with the expression in Eq. (2.6), since we are primarily interested in the physics near $\pm K^*$ on the FS. The bosons that couple most strongly to those two regions carry momenta satisfying $k_{d-m} \ll |\mathbf{L}_{(k)}|$, so that $\sqrt{k_{d-m}^2 + \cdots + k_d^2}$ reduces to $|\mathbf{L}_{(k)}|$ to good approximation.

Incorporating the dressed bosonic propagator, $D_1$, the one-loop fermionic self-energy, $\Sigma_1(q)$, needs to be computed, which is shown in Fig. 2.4b. An explicit computation leads to [9]

$$\Sigma_1(q) = \frac{(i\,e)^2\,\mu^{x_e}}{N}\int dk\,\gamma_{d-m}\,G_0(q-k)\,\gamma_{d-m}\,D_1(k)$$
$$= \frac{-\,i\,e^{2\,(m+1)/3}\,\mu^{x_e\,(m+1)/3}\,\mathbf{\Gamma}\cdot\mathbf{Q}}{6\,N\,\pi^{(m-1)/2}(4\pi)^{\frac{d-m}{2}}\,2^{m-1}\,|\sin\{(m+1)\pi/3\}|\,\beta_d^{(2-m)/3}\,(\mu\,\tilde{k}_F)^{(m-1)(2-m)/6}}$$
$$\times\,\frac{\Gamma(\frac{3-(m+1)(d-m)}{6})\,\Gamma(\frac{d-m-2\beta}{2})\,\Gamma(\frac{d-m+1}{2})}{\Gamma(m/2)\,\Gamma(\beta)\,\Gamma(d-m-\beta+\frac{1}{2})\,(\mathbf{Q}^2)^{\frac{3-(m+1)(d-m)}{6}}}, \tag{2.7}$$

where $\beta \equiv (d-m)\,(2-m)/6$. $\Sigma_1(q)$ shows a logarithmic divergence in the IR-scale $\Lambda$ at $d_c(m) = m + 3/(m+1)$. The value of $d_c(m)$ is clearly smaller than $\tilde{d}_c$ within the range $1 < m < 2$. Setting $d = d_c(m) - \epsilon$, the divergence $\sim \ln\Lambda$ corresponds to the divergence captured by the pole, $1/\epsilon$, when translated into the language of dimensional regularization. This can be seen by rewriting the fermionic self-energy as

$$\Sigma_1(q) = \left[-\frac{e^{2\,(m+1)/3}}{\tilde{k}_F^{\frac{(m-1)(2-m)}{6}}}\frac{u_1}{N\,\epsilon} + \text{finite terms}\right](i\,\mathbf{\Gamma}\cdot\mathbf{Q}),$$
$$u_1 = \frac{(\pi/\beta_d)^{\frac{2-m}{2}}}{(4\pi)^{\frac{3}{2(m+1)}}\,2^{m-1}\left|\sin\left(\frac{(m+1)\pi}{3}\right)\right|}\,\frac{\Gamma\left(\frac{m+4}{2(m+1)}\right)}{\Gamma(\frac{m}{2})\,\Gamma\left(\frac{2-m}{2(m+1)}\right)\Gamma\left(\frac{2m+5}{2(m+1)}\right)}, \tag{2.8}$$

keeping terms upto the leading order in $q/k_F$. The one-loop vertex correction is shown in the Feynman diagram of Fig. 2.4c. It vanishes identically, which can be traced back to the existence of a Ward identity [3]. The key advantage of our formalism is that we can tune the value of $m$ from 1 to 2 with the help of the expansion-parameter $\epsilon$, which remains small as required. Thus, we achieve a controlled perturbative expansion through $\epsilon$, as long as we remain within $1 \le m \le 2$. We would like to emphasize that we are tuning the two dimensionalities, namely, $m$ and $d$, independently, which of course must obey the constraint $m < d$ to make any physical sense. This allows us to tune $d$ such that $\epsilon = d_c(m) - d$ is perturbatively small for a given $m$.

Equation (2.5) is taken to be the *physical action*, defined at an energy scale $\mu \sim \Lambda$, and consists of the fundamental Lagrangian expressed in terms of non-divergent quantities. However, loop integrals generically produce divergent contributions, and to handle these we employ renormalization via dimensional regularization. The UV-divergent terms are those that arise in the $\epsilon \to 0$ limit. To systematically control these divergences, we work within the minimal subtraction (MS) renormalization scheme [16, 17], in which the divergent parts of loop contributions are cancelled by adding appropriate counterterms. More precisely, we adopt the modified minimal subtraction ($\overline{\text{MS}}$) scheme, wherein one absorbs not only the strictly divergent pole but also the universal finite term proportional to $\epsilon^0$ that invariably accompanies the $1/\epsilon$ pole into the corresponding counterterm.

The counterterm action, $S_{CT}$, which is designed to absorb all singular contributions, is constructed by introducing counterterm factors as power series $A_\zeta = \sum_{n=1}^{\infty} Z_{\zeta,n}/\epsilon^n$ with $\zeta \in [1, 4]$, chosen so as to cancel the divergent $1/\epsilon^n$ contributions from the Feynman diagrams. It takes the form of

$$\begin{aligned} S_{CT} = & \sum_\lambda \int dk\, \bar{\Psi}_\lambda(k)\, i \Big[ A_1\, \mathbf{\Gamma} \cdot \mathbf{K} + A_2\, \gamma_{d-m}\, \delta_k \Big] \Psi_\lambda(k)\, e^{\frac{\mathbf{L}^2_{(k)}}{\mu k_F}} \\ & + \frac{A_3}{2} \int dk\, \mathbf{L}^2_{(k)}\, \phi(-k)\, \phi(k) \\ & + \frac{i\, e\, \mu^{x_e/2}}{\sqrt{N}} \sum_\lambda \int dkdq\; A_4\, \phi(q)\, \bar{\Psi}_\lambda(k+q)\, \gamma_{d-m} \Psi_\lambda(k). \end{aligned} \tag{2.9}$$

Due to the $(d-1)$-dimensional rotational invariance in the space perpendicular to the FS, each term in $\mathbf{\Gamma} \cdot \mathbf{K}$ is renormalized in the same way. A Ward identity enforces $A_4 = A_3$ [3]. Subtracting $S_{CT}$ from the so-called *bare* action $S_{\text{bare}}$, we obtain the renormalized action, which serves as the *physical* effective action of the theory, rewritten entirely in terms of non-divergent quantum parameters. Here,

$$\begin{aligned} S_{\text{bare}} = & \sum_\lambda \int dk_B\, \bar{\Psi}_{B\lambda}(k_B)\, i \Big[ \mathbf{\Gamma} \cdot \mathbf{K}_B + \gamma_{d-m} \delta_{k_B} \Big] \Psi_{B\lambda}(k_B)\, \exp\left\{ \frac{\mathbf{L}^2_{(k),B}}{k_{F,B}} \right\} \\ & + \frac{1}{2} \int dk_B\, \mathbf{L}^2_{(k)}\, \phi_B(-k_B)\, \phi_B(k_B) \\ & + \frac{i\, e_B}{\sqrt{N}} \sum_\lambda \int dk_B\, dq_B\, \phi_B(q_B)\, \bar{\Psi}_{B\lambda}(k_B + q_B)\, \gamma_{d-m} \Psi_{B\lambda}(k_B), \end{aligned} \tag{2.10}$$

consisting of the *bare quantities*, each labelled with the subscript "$B$". While the bare parameters may be divergent, the physical observables are identified with the renormalized coupling constants, which evolve as functions of the logarithm of the floating energy scale $\mu = \mu_0\, e^{-l}$. We now relate the bare quantities to their renormalized counterparts—written without the superscript "$B$"—via the multiplicative $Z_\zeta$-factors, such that $S_{\text{bare}} = S + S_{CT}$ and $Z_\zeta = 1 + A_\zeta$. Here,

$$\begin{aligned} & \mathbf{K} = \frac{Z_2}{Z_1} \mathbf{K}_B, \quad k_{d-m} = k_{B,d-m}, \quad \mathbf{L}_{(k)} = \mathbf{L}_{(k),B}, \quad \Psi(k) = \frac{\Psi_B(k_B)}{\sqrt{Z_\Psi}}, \quad k_F = \mu\, \tilde{k}_F, \\ & \phi(k) = \frac{\phi_B(k_B)}{\sqrt{Z_\phi}}, \quad e_B = \frac{1}{\sqrt{Z_3}} \left( \frac{Z_2}{Z_1} \right)^{\frac{d-m}{2}} \mu^{\frac{x_e}{2}}\, e, \\ & Z_\Psi = Z_2 \left( \frac{Z_2}{Z_1} \right)^{d-m}, \quad Z_\phi = Z_3 \left( \frac{Z_2}{Z_1} \right)^{d-m}. \end{aligned}$$

Restricting to the one-loop order, we have $Z_\zeta = 1 + \frac{Z_{\zeta,1}}{\epsilon}$, with

$$Z_{1,1} = -\, e^{2\,(m+1)/3}\, u_1 / \big[\tilde{k}_F^{\frac{(m-1)\,(2-m)}{6}}\, N\big] \text{ and } Z_{2,1} = Z_{3,1} = Z_{4,1} = 0. \tag{2.11}$$

In the language of Renormalization Group (RG) transformations, the beta-function of a coupling-constant $g$ is defined as $\beta_g = \partial_{\ln \mu} g = \mu\, \partial_\mu g \equiv \partial_l g$. As we integrate out energy-shells, $\beta_g$ is the function that determines the flow of $g$ with $\ln \mu$. At one-loop level, for the two coupling-constants, $\tilde{k}_F$ and $e$, the beta-functions take the forms of

$$\beta_{\tilde{k}_F} = -\,\tilde{k}_F, \quad \beta_e = -\left[\frac{\epsilon}{2} + \frac{(m-1)\,(2-m)}{4\,(m+1)}\right] e + \frac{u_1\, \tilde{e}}{2\,N}\, e,$$
$$\tilde{e} \equiv \frac{e^{2(m+1)/3}}{\tilde{k}_F^{\frac{(m-1)\,(2-m)}{6}}}, \quad z = 1 - \frac{\partial}{\partial \ln \mu}\left(\frac{Z_2}{Z_1}\right), \quad \eta_\psi = \frac{\mu}{2}\frac{\partial Z_\psi}{\partial \mu}, \quad \eta_\phi = \frac{\mu}{2}\frac{\partial Z_\phi}{\partial \mu}. \tag{2.12}$$

Here, $z$ is the dynamical critical exponent and $\eta_\psi$ ($\eta_\phi$) incorporates the anomalous dimension for the fermions (bosons). The quantity $\tilde{k}_F$ increases under the RG flow because the size of the FS, measured in units of $\mu$, grows at low energies. The first term in $\beta_e$ indicates that $e$ remains strictly relevant at $d = d_c(m)$ for $1 < m < 2$, as reflected by $x_e > 0$ and $\tilde{d}_c > d_c$. Intriguingly, the form of the loop correction—the second term—reveals that higher-order corrections are controlled not by $e$ itself, but by an effective coupling $\tilde{e}$, which is responsible for the $e^{fr}$ powers that accompany them, as discussed earlier. The beta-function for $\tilde{e}$, which no longer contains $\tilde{k}_F$, is given by

$$\beta_{\tilde{e}} = -\,\frac{(m+1)\,\epsilon}{3}\,\tilde{e} + \frac{(m+1)\,u_1}{3N}\,\tilde{e}^2 + \mathcal{O}(\tilde{e}^3). \tag{2.13}$$

From this, we see that $\tilde{e}$ flows to an IR-stable fixed point at $\tilde{e}^* = N\epsilon/u_1 + \mathcal{O}(\epsilon^2)$. For small $\epsilon$, this interacting fixed point is perturbatively accessible, despite the fact that the scaling dimension $x_e$ of the bare coupling $e$ remains positive in the $\epsilon \to 0$ limit for $1 < m < 2$. Although $e$ grows at low energies, this growth is compensated by the Landau damping, which also strengthens as the effective size of the FS increases. It is the competition between the interaction and the Landau damping that renders the effective coupling marginal at the true critical dimension $d_c$, which does not coincide with $\tilde{d}_c$ for $1 < m < 2$. We also observe that $k_F$ drops out of the effective coupling not only for $m = 1$ but also for $m = 2$. In the latter case, the $k_F$ dependence arising from the Landau damping is precisely cancelled by the $k_F$ dependence from the phase space of intermediate states in Fig. 2.4b. Nevertheless, UV/IR mixing persists for all $m > 1$, since the Landau damping diverges in the large-$k_F$ limit. Finally, the fixed-point values of $\{z, \eta_\psi, \eta_\phi\}$ evaluate at one-loop order to $z^* = 3/[3 - (m+1)\,\epsilon]$ and $\eta_\psi^* = \eta_\phi^* = -\epsilon/2$. It is remarkable that these exponents turn out to be insensitive to the details of the FS—such as $\beta_d$—despite the fact that patch scaling is violated by $k_F$. This vindicates our use of the exponential cutoff scheme in Eq. (2.3), which captures the compactness of the FS in a minimal way without requiring knowledge of its precise shape. The finite anomalous dimensions arise from a dynamical balance between the two strongly relevant couplings, $e$ and $k_F$—a mechanism that is qualitatively opposite to cases where finite anomalous dimensions emerge from a balance between two *irrelevant couplings* [7].

## 2.4 Renormalization-Group Equations

A renormalized Green's function, $G^{(n_\psi,n_\psi,n_\phi)}$, defined by the expectation value,

$$\left\langle \phi(k^1)\cdots\phi(k^{n_\phi})\,\Psi(k^{n_\phi+1})\cdots\Psi(k^{n_\phi+n_\psi})\,\bar{\Psi}(k^{n_\phi+n_\psi+1})\cdots\bar{\Psi}(k^{n_\phi+2n_\psi})\right\rangle$$

$= G^{(n_\psi,n_\psi,n_\phi)}(\{k^\alpha\};\tilde{e},\tilde{k}_F,\mu)\,\delta^{d+1}\left(\sum_{\alpha=1}^{n_\phi+n_\psi}k^\alpha-\sum_{\alpha'=n_\phi+n_\psi+1}^{2n_\psi+n_\phi}k^{\alpha'}\right)$, is the one which satisfies the RG equation. This is expressed as

$$\Bigg[-\sum_{\alpha=1}^{2n_\psi+n_\phi}\left(z\,\mathbf{K}^\alpha\cdot\nabla_{K^\alpha}+k^{\alpha'}_{d-m}\,\partial_{k^{\alpha'}_{d-m}}+\frac{\mathbf{L}_{(k^\alpha)}}{2}\cdot\nabla_{L_{(k^\alpha)}}\right)+\beta_{\tilde{k}_F}\partial_{\tilde{k}_F}+\beta_{\tilde{e}}\partial_{\tilde{e}}$$
$$+2\,n_\psi\left(-\frac{2\,d_c-2\epsilon+4-m}{4}+\eta_\psi\right)+n_\phi\left(-\frac{2\,d_c-2\epsilon+4-m}{4}+\eta_\phi\right)+d_c-\epsilon$$
$$+1-\frac{m}{2}+(d_c-\epsilon-m)\,(z-1)\Bigg]\,G^{(n_\psi,n_\psi,n_\phi)}(\{k_i\};\tilde{e},\tilde{k}_F,\mu)=0. \tag{2.14}$$

A generic $G^{(n_\psi,n_\psi,n_\phi)}$ contains both fermionc and bosonic fields. In particular, we are interested in the two-point functions for the bosons and the fermions because they contain the information of the generic scaling-form at the IR fixed point. These are:

$$\begin{aligned}
G^{(0,0,2)} &= \frac{1}{\left(\mathbf{L}^2_{(k)}\right)^{2\Delta_\phi}}\,f_F\left(\frac{|\mathbf{K}|^{1/z^*}}{\mathbf{L}^2_{(k)}},\,\frac{k_{d-m}}{k_F},\,\frac{\mathbf{L}^2_{(k)}}{k_F}\right),\\
G^{(1,1,0)} &= \frac{1}{|\delta_k|^{2\Delta_\psi}}\,f_B\left(\frac{|\mathbf{K}|^{1/z^*}}{\delta_k},\,\frac{\delta_k}{k_F},\,\frac{\mathbf{L}^2_{(k)}}{k_F}\right),
\end{aligned} \tag{2.15}$$

where $2\,\Delta_\phi=1-(z^*-1)\left(\frac{3}{m+1}-\epsilon\right)-2\,\eta^*_\phi=1+O(\epsilon^2)$ and $2\,\Delta_\psi=1-(z^*-1)\left(\frac{3}{m+1}-\epsilon\right)-2\,\eta^*_\psi=1+O(\epsilon^2)$. From the expressions at one-loop order, we infer the universal scaling-forms to be

$$f_B(X,Y,Z) = \left[1 + \beta_d\, \tilde{e}^{\frac{3}{m+1}}\, X^{\frac{3}{m+1}}\, Z^{-\frac{3(m-1)}{2(m+1)}}\right]^{-1},$$
$$f_F(X,Y,Z) = -i\left[C\,(\mathbf{\Gamma}\cdot\hat{\mathbf{K}})\,X + \gamma_{d-m}\right]^{-1},\; C = \mu^{\frac{m+1}{3}\epsilon}\left[1 - \frac{(m+1)\,\gamma\,\epsilon}{6}\right], \tag{2.16}$$

in the limit $Y, Z \to 0$ with $X$ held fixed. The expressions show that $f_D$ has a singular dependence on $Z$ in the $Z \to 0$ limit. We notice that the sliding symmetry being absent for the momentum coordinates of $\delta_k$ and $\mathbf{L}_{(k)}$, the fermionic Green's function's dependence on $\delta_k$ and $\mathbf{L}_{(k)}$ is distinct in general.

## 2.5 Physical Relevance of the Expansion-Parameter $\epsilon$

The motivation for employing dimensional regularization and, ultimately, the small-$\epsilon$ expansion is to shed light on the stark difference in the characteristics of NFLs with $d_p = 2$ and $d_p = 3$. Since the former has an $m = 1$ FS, $k_F$ plays no role in the low-energy scaling. By contrast, $k_F$ enters as an important scale for $d_p = 3$ NFLs precisely due to UV/IR mixing. We now examine how this transition occurs in a systematic way by tuning $d$ and $m$ continuously. Although systems with non-integer dimensions are unphysical in their own right, they provide valuable insight into how $d$ and $m$ each contribute to the distinct properties of metals in actual physical dimensions.

For $d_p = 3$, one has $m = 2$, and $k_F$ is seen to drop out of the expression for $\tilde{e}$. Nevertheless, the intrinsic UV/IR mixing still manifests itself in the dispersion relations of the fermion and boson, which go as $k_0 \sim k_x + \mathbf{L}_{(k)}^2$ and $k_0 \sim \mathbf{L}_{(k)}^3$, respectively, up to small corrections. These two fields can have different effective dynamical critical exponents at the scale-invariant fixed point, with the mismatch compensated by $k_F$. Our analysis therefore yields the correct scaling behavior, consistent simultaneously for both bosons and fermions, by incorporating the dimensional parameter $k_F$ in an appropriate way. This stands in contrast to the $m = 1$ case, where the dispersions of bosons and fermions obey the same scaling behavior [3]. UV/IR mixing also plays an important role in suppressing higher-loop quantum corrections for $m > 1$, a point we turn to in the next section.

## 2.6 Higher-Loop Diagrams And Expansion Parameter

In Ref. [3], the $d_p = 2$ case was treated, where the two- and three-loop Feynman diagrams were shown to be suppressed by positive powers of $\tilde{e}$, with $\tilde{e} = \mathcal{O}(\epsilon)$. A general argument also exists outlining why higher-loop diagrams are systematically suppressed by successively higher powers of $\tilde{e}$. Our systematic expansion in $\epsilon$ is distinct from an expansion in powers of $1/N$, and does not suffer from the proliferation of planar diagrams that afflicts the $1/N$ expansion [11, 18]. The introduction of extra codimensions mathematically suppresses the density of states at low energies, thereby reducing quantum fluctuations. This results in a weaker infrared singularity, which in turn permits a controlled expansion for sufficiently small $\epsilon$. We have generalized the treatment of Ref. [3] to $d_p > 2$, where consequently $m > 1$. While the suppression of higher-loop diagrams by positive powers of $\tilde{e}$ persists unchanged for $m > 1$, a qualitatively new feature emerges: the explicit appearance of an additional energy-scale in the form of $k_F$.

To make an estimate of the magnitude of higher-loop corrections, we first discuss an interplay between $k_F$ and $\Lambda$ that plays an important role for $m > 1$. Let $k = \{\mathbf{K}, k_{d-m}, \mathbf{L}_{(k)}\}$ denote the momentum flowing through a boson propagator within a two-loop or higher-loop diagram. When $|\mathbf{K}|$ is of order $\Lambda$, the typical momentum carried by the boson along the tangential direction of the FS is given by $|\mathbf{L}_{(k)}|^3 \sim \tilde{\alpha}\, \Lambda^{d-m}$, where $\tilde{\alpha} = \beta_d\, e^2\, \mu^{x_e}\, (\mu\, \tilde{k}_F)^{(m-1)/2}$ [cf. Eq. (2.6)]. When $(\tilde{\alpha}\, \Lambda^{d-m})^{1/3} \gg \sqrt{\Lambda}$, the momentum imparted from the boson to the fermion greatly exceeds $\sqrt{\Lambda}$, as illustrated in Fig. 2.3b. Physically, this means that the typical energy of virtual particle-hole excitations within the loop is much larger than $\Lambda$, and as a result the loop contributions are suppressed by a power of $\Lambda/k_F$ at low energies. By contrast, no such suppression arises when $(\tilde{\alpha}\, \Lambda^{d-m})^{1/3} \ll \sqrt{\Lambda}$. The crossover between these two regimes is controlled by the dimensionless quantity $\lambda_{\text{cross}} \equiv \tilde{e}^2 (k_F/\Lambda)^{m-1}$, which determines whether $(\tilde{\alpha}\, \Lambda^{d-m})^{1/3}$ is much greater or much less than $\Lambda^{1/2}$.

For $m = 1$, the $k_F$ dependence drops out entirely, including from the higher-loop diagrams. Since $\tilde{e} \sim \mathcal{O}(\epsilon)$ within the perturbative window, one always operates in the limit $\lambda_{\text{cross}} \ll 1$ for $m = 1$. The situation is qualitatively different for $m > 1$, where the tangential momentum carried by the boson depends on both $\Lambda$ and $k_F$. Indeed, for a fixed value of $\tilde{e} \sim \mathcal{O}(\epsilon)$, one is always driven into the regime $\lambda_{\text{cross}} \gg 1$ for $m > 1$ at sufficiently low energies, since $k_F$ carries a positive scaling dimension and $k_F/\Lambda$ flows to infinity in the low-energy limit. The crossover occurs at the energy scale $\Lambda \sim \tilde{e}^{2/(m-1)}\, k_F$. It is worth noting that for small $\epsilon$ and $(m - 1)$, there exists a wide energy window before the theory enters the low-energy regime controlled by $\lambda_{\text{cross}} \gg 1$. In this strictly low-energy limit, higher-loop diagrams are suppressed by $k_F$, as has also been pointed out in Ref. [19] for the specific case of $d_p = 3$ and $m = 2$.

For general $m > 1$ with $\lambda_{\text{cross}} \gg 1$, the two-loop bosonic and fermionic self-energies are given by [9]

$$\Pi_2(k) \sim \frac{\tilde{e}^{\frac{m}{m+1}}}{k_F^{\frac{m-1}{2(m+1)}}} \frac{|\mathbf{K}|^{\frac{m}{m+1}}}{N\,|\mathbf{L}_{(k)}|}\, \Pi_1(q) \quad \text{and} \quad \Sigma_{2a}(k) \sim \tilde{e}^{\frac{2(m-1)}{m+1}} \left(\frac{\Lambda}{k_F}\right)^{\frac{2(m-1)}{m+1}} i\, \gamma_{d-m}\, \delta_k,$$

respectively, to leading order in $\Lambda/k_F$. The vertex correction is related to the fermionic self-energy through the Ward identity. Compared to the one-loop self-energies, the two-loop corrections are therefore suppressed not only by $\tilde{e}$ but also by powers of $\Lambda/k_F$. Owing to this suppression by $1/k_F$, no logarithmic or higher-order divergence appears at the critical dimension, and consequently the critical exponents receive no corrections from the two-loop diagrams in the $k_F \to \infty$ limit. The suppression by $\Lambda / k_F$ can be traced back to the large Landau damping, which quenches quantum fluctuations at low energies. Since this mechanism is not specific to the two-loop diagrams, we expect all higher-loop contributions to be suppressed by $\tilde{e}$ and $1/k_F$ in the low-energy limit—a conclusion we have verified explicitly for the Aslamazov-Larkin-type diagrams comprising three loops.

## 2.7 Outlook

In this chapter, we have demonstrated how to extract the scaling behaviour of NFLs with a FS of general dimensions and co-dimensions, using a dimensional regularization scheme. For $m > 1$, the low-energy physics becomes sensitive to the size of FS, $k_F$, which results in an intriguing phenomenon of UV/IR mixing. By tuning the dimension below the upper critical dimension, we have shown that there exists a stable NFL fixed point where both interaction and UV/IR mixing play crucial roles. We have also shown that the critical exponents at the low-energy fixed point are not modified by higher-loop diagrams for $m > 1$. In the next chapter, we will demonstrate how the Ising-nematic order parameter provides a "stronger glue" compared to phonons, bringing about unconventional superconductivity.

## References


1. S. Sachdev, *Quantum Phase Transitions*, 2nd edn. (Cambridge University Press, 2011)
2. S.S. Lee, Annu. Rev. Condens. Matter Phys. **9**, 227 (2018). https://doi.org/10.1146/annurev-conmatphys-031016-025531
3. D. Dalidovich, S.S. Lee, Phys. Rev. B **88**, 245106 (2013). https://doi.org/10.1103/PhysRevB.88.245106
4. A.L. Fitzpatrick, S. Kachru, J. Kaplan, S. Raghu, Phys. Rev. B **88**, 125116 (2013). https://doi.org/10.1103/PhysRevB.88.125116. http://link.aps.org/doi/10.1103/PhysRevB.88.125116
5. G. Torroba, H. Wang, Phys. Rev. B **90**, 165144 (2014). https://doi.org/10.1103/PhysRevB.90.165144. http://link.aps.org/doi/10.1103/PhysRevB.90.165144
6. S. Chakravarty, R.E. Norton, O.F. Syljuåsen, Phys. Rev. Lett. **74**, 1423 (1995). https://doi.org/10.1103/PhysRevLett.74.1423
7. S. Sur, S.S. Lee, Phys. Rev. B **91**, 125136 (2015). https://doi.org/10.1103/PhysRevB.91.125136
8. T. Senthil, R. Shankar, Phys. Rev. Lett. **102**, 046406 (2009). https://doi.org/10.1103/PhysRevLett.102.046406
9. I. Mandal, S.S. Lee, Phys. Rev. B **92**, 035141 (2015). https://doi.org/10.1103/PhysRevB.92.035141
10. M.A. Metlitski, S. Sachdev, Phys. Rev. B **82**, 075127 (2010). https://doi.org/10.1103/PhysRevB.82.075127
11. M.A. Metlitski, S. Sachdev, Phys. Rev. B **82**, 075128 (2010). https://doi.org/10.1103/PhysRevB.82.075128
12. I. Mandal, Eur. Phys. J. B **89**(12), 278 (2016). https://doi.org/10.1140/epjb/e2016-70509-4
13. D. Pimenov, I. Mandal, F. Piazza, M. Punk, Phys. Rev. B **98**, 024510 (2018). https://doi.org/10.1103/PhysRevB.98.024510

14. I. Mandal, Nucl. Phys. B **1005**, 116586 (2024). https://doi.org/10.1016/j.nuclphysb.2024.116586
15. I. Mandal, Ann. Phys. **474**, 169925 (2025). https://doi.org/10.1016/j.aop.2025.169925
16. G. 't Hooft, Nucl. Phys. B **61**, 455 (1973). https://doi.org/10.1016/0550-3213(73)90376-3
17. S. Weinberg, Phys. Rev. D **8**, 3497 (1973). https://doi.org/10.1103/PhysRevD.8.3497
18. S.S. Lee, Phys. Rev. B **80**, 165102 (2009). https://doi.org/10.1103/PhysRevB.80.165102
19. T. Schäfer, K. Schwenzer, Phys. Rev. D **70**, 054007 (2004). https://doi.org/10.1103/PhysRevD.70.054007

# Chapter 3
# Enhancement of Superconductivity by Ising-Nematic Order Parameter

## 3.1 Introduction

We continue our exploration of the Ising-nematic ordering [1, 2] by examining its effect on a four-fermion interaction of strength $V$ in the pairing channel. Taking a system embedded in $d$ spatial dimensions (see Chap. 1) possessing an $m$-dimensional Fermi surface (FS), the tree-level scaling dimension of $V$ is $[V] = -d + 1 + m/2$. As we shall see, scatterings in the BCS channel are enhanced by powers of the FS volume, $\sim k_F^m$, so that the effective coupling governing the potential superconducting instability becomes $\tilde{V} = V\, k_F^{m/2}$, with an enhanced scaling dimension of $[\tilde{V}] = -d + 1 + m$. This effective coupling is marginal precisely when the co-dimension takes the value $d - m = 1$. Our goal is to understand how the interactions between the fermions and the critical bosons conspire to enhance a pairing instability [3].

## 3.2 Superconducting Instability

To investigate the emergence of superconducting instabilities within the prescribed non-Fermi liquid framework, we augment the fermion-boson action introduced in Chap. 1 with the requisite four-fermion interaction terms. The following terms can give rise to Cooper pairing in the BCS channel:



I. Mandal, *Controlled Description Of Non-Fermi Liquids Using Field-Theoretic Methods*, SpringerBriefs in Physics, https://doi.org/10.1007/978-3-032-21618-2_3

$$\int \left(\prod_{s=1}^{4} dp_s\right) \frac{\mathcal{F}(p_1, p_3; p_2, p_4)}{4} \Big[\{\bar{\Psi}_{j_1}(p_3)\,\Psi_{j_2}(p_1)\}\,\{\bar{\Psi}_{j_3}(p_4)\,\Psi_{j_4}(p_2)\}$$
$$- \{\bar{\Psi}_{j_3}(p_3)\sigma_z\Psi_{j_4}(p_1)\}\,\{\bar{\Psi}_{j_1}(p_4)\sigma_z\Psi_{j_2}(p_2)\}\Big]$$
$$= - \int \left(\prod_{s=1}^{4} dp_s\right) \mathcal{F}(p_1, p_3; p_2, p_4)\Big[\psi^\dagger_{+,j_1}(p_3)\,\psi^\dagger_{-,j_2}(-p_1)\,\psi_{-,j_3}(-p_4)\,\psi_{+,j_4}(p_2)\Big].$$

In this context, the vertex function $\mathcal{F}(p_1, p_3; p_2, p_4)$ exhibits invariance under the simultaneous exchange of indices $(p_1, p_3) \leftrightarrow (p_2, p_4)$. Consequently, the action defined in Chap. 1 is supplemented by

$$S^{\rm SC}_{\rm gen} = \frac{\mu^{d_v}}{4} \sum_{j_1, j_2, j_3, j_4} \int \left(\prod_{s=1}^{4} dp_s\right) (2\pi)^{d+1}\,\delta^{(d+1)}(p_1 + p_2 - p_3 - p_4)$$
$$\times \Big[\{\bar{\Psi}_{j_1}(p_3)\Psi_{j_2}(p_1)\}\,\{\bar{\Psi}_{j_3}(p_4)\Psi_{j_4}(p_2)\}$$
$$- \{\bar{\Psi}_{j_3}(p_3)\sigma_z\Psi_{j_4}(p_1)\}\,\{\bar{\Psi}_{j_1}(p_4)\sigma_z\Psi_{j_2}(p_2)\}\Big]$$
$$\times \Big[V_S(\mathbf{p}_1, \mathbf{p}_3; \mathbf{p}_2, \mathbf{p}_4)\left(\delta_{j_1,j_3}\delta_{j_2,j_4} - \delta_{j_1,j_4}\delta_{j_2,j_3}\right)$$
$$+ V_A(\mathbf{p}_1, \mathbf{p}_3; \mathbf{p}_2, \mathbf{p}_4)\left(\delta_{j_1,j_3}\delta_{j_2,j_4} + \delta_{j_1,j_4}\delta_{j_2,j_3}\right)\Big],$$

The subscripts $S$ and $A$ denote that the vertex functions $V_S(\mathbf{p}_1, \mathbf{p}_3; \mathbf{p}_2, \mathbf{p}_4)$ and $V_A(\mathbf{p}_1, \mathbf{p}_3; \mathbf{p}_2, \mathbf{p}_4)$ possess symmetric and antisymmetric properties, respectively, under the interchange $p_1 \leftrightarrow p_3$ or $p_2 \leftrightarrow p_4$. These couplings are rendered dimensionless via the mass scale $\mu$, where $d_v = -d + 1 + \frac{m}{2}$ characterizes the scaling dimension of the interaction. A Renormalization group (RG) analysis will reveal that a superconducting instability necessitates the kinematic constraint $\mathbf{p}_1 = \mathbf{p}_3$ and $\mathbf{p}_2 = \mathbf{p}_4$. Furthermore, the assumption of rotational invariance permits the simplification $V_{S/A}(\mathbf{p}_1, \mathbf{p}_1; \mathbf{p}_2, \mathbf{p}_2) = V_{S/A}(\theta_1 - \theta_2)$, where $\theta_1$ and $\theta_2$ parameterize the angular positions on the Fermi surface. To facilitate an analytical treatment, we restrict our focus to the $s$-wave channel, characterized by a constant non-zero $V_S$. This channel requires a minimum of two fermion flavors; accordingly, for $N = 2$, the effective action reduces to

$$S^{\rm SC} = \frac{-\mu^{d_v} V_S}{4} \sum_{j_1, j_2} \int \left(\prod_{s=1}^{4} dp_s\right) (2\pi)^{d+1}\,\delta^{(d+1)}(p_1 + p_2 - p_3 - p_4)\left(1 - \delta_{j_1,j_2}\right)$$
$$\times \Big[\{\bar{\Psi}_{j_1}(p_3)\Psi_{j_2}(p_1)\}\,\{\bar{\Psi}_{j_2}(p_4)\Psi_{j_1}(p_2)\}$$
$$- \{\bar{\Psi}_{j_1}(p_3)\sigma_z\Psi_{j_2}(p_1)\}\,\{\bar{\Psi}_{j_2}(p_4)\sigma_z\Psi_{j_2}(p_1)\}\Big]. \tag{3.1}$$

The results from this simple treatment can be readily generalized to systems characterized by an arbitrary number of fermion flavours, $N > 2$, and superconducting instability involving nonzero values of angular momentum.

### 3.2.1 One-Loop Diagrams Generating Terms Proportional to $V_S^2$

To account for contributions proportional to $V_S^2$, we evaluate the one-loop diagrams illustrated in Fig. 3.2. The contribution from Fig. 3.2a is determined to be proportional to

$$
\begin{aligned}
& -(1-\delta_{j_1,j_2})^2 \int dk \, \mathrm{Tr}\big[G_0(k+\mathbf{P_1}-\mathbf{P_3})\, G_0(k)\big] \\
&= \frac{2^{1-2d+\frac{m}{2}} \left(1-\delta_{j_1,j_2}\right) k_F^{m/2}\, \pi^{1-\frac{d}{2}} \sec\left(\frac{(d-m)\,\pi}{2}\right)}{\Gamma\left(\frac{d-m}{2}\right) |\mathbf{P_3}-\mathbf{P_1}|^{-d+m+1}},
\end{aligned}
$$

which is logarithmically-divergent at $d-m=1$. Hence, we express it as $\frac{k_F^{m/2} \ln\left(\frac{\Lambda}{|\mathbf{P_1}-\mathbf{P_3}|}\right)}{2^{\frac{3m}{2}}\, \pi^{1+\frac{m}{2}}}$. The results from Fig. 3.2b and c are suppressed by powers of $k_F$ and, hence, do not contribute to the beta-functions. Noting that $\mathrm{Tr}\big[\sigma_z\, G_0(k+\mathbf{P_1}-\mathbf{P_3})\, \sigma_z\, G_0(k)\big] = -\mathrm{Tr}\big[G_0(k+\mathbf{P_1}-\mathbf{P_3})\, G_0(k)\big]$, the full contribution from all one-loop diagrams proportional to $V_S^2$ is given by

$$
\begin{aligned}
t_{VV} &= \frac{4\times 2}{2!}\, \frac{4}{4\times 4} \times \frac{2^{2+2d-\frac{m}{2}}\, k_F^{m/2}\, \mu^{2d_v}\, V_S^2}{\pi^{\frac{d}{2}}\, \Gamma\left(\frac{d-m}{2}\right) \epsilon} + \mathcal{O}(\epsilon) = \frac{v_2\, \mu^{2d_v+\frac{m}{2}}\, \tilde{k}_F^{m/2}\, V_S^2}{\epsilon} + \mathcal{O}(\epsilon), \\
&\text{where } v_2 = \frac{2^{2-2d+\frac{m}{2}}}{\pi^{\frac{d}{2}}\, \Gamma\left(\frac{d-m}{2}\right)} \text{ and } d-m = 1-\epsilon.
\end{aligned}
\tag{3.2}
$$

In accordance with previous treatments in the literature [4, 5], the analysis must incorporate tree-level contributions within the BCS channel, arising from long-range interactions between fermions. In the present context, these interactions are mediated by the exchange of massless bosons. Consistent with our treatment of competing terms, these contributions are evaluated using the formalism of dimensional regularization. Specifically, the diagrams in Fig. 3.1 yield the terms,

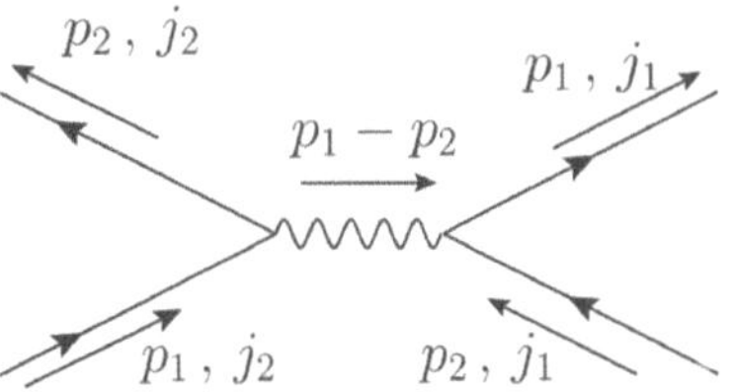


**Fig. 3.1** Tree-level Feynman diagram proportional to $e^2$

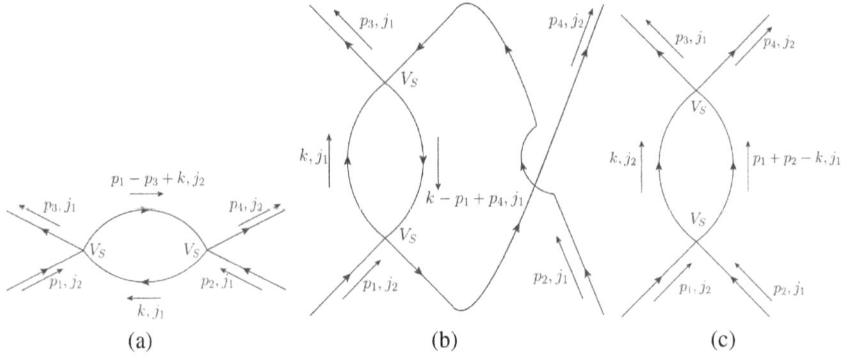


**Fig. 3.2** One-loop diagrams proportional to $V_S^2$. Here, $p_4 = p_1 + p_2 - p_3$

$$
\begin{aligned}
\sum_{j_1, j_2} \Big[ & - \psi^\dagger_{+,j_1}(p_2)\, \psi^\dagger_{-,j_2}(-p_2)\, \psi_{-,j_2}(-p_1)\, \psi_{+,j_1}(p_1) \\
& - \psi^\dagger_{+,j_1}(p_1)\, \psi^\dagger_{-,j_2}(-p_1)\, \psi_{-,j_2}(-p_2)\, \psi_{+,j_1}(p_2) \\
& - \psi^\dagger_{+,j_1}(p_2)\, \psi_{+,j_1}(p_1)\, \psi^\dagger_{+,j_2}(p_1)\, \psi_{+,j_2}(p_2) \\
& - \psi^\dagger_{-,j_2}(-p_2)\, \psi_{-,j_2}(-p_1)\, \psi^\dagger_{-,j_1}(-p_1)\, \psi_{-,j_1}(-p_2) \Big]
\end{aligned}
$$

which multiply $\frac{(i\,e)^2 \mu^{x_e} D_1(p_1 - p_2)}{2N}$. Owing to the symmetry of $D_1(p_1 - p_2)$ under the interchange $p_1 \leftrightarrow p_2$, this term contributes as $\frac{e^2 \mu^{x_e} D_1(p_1 - p_2)}{N}$ to the components governing the pairing instability.

Using $\mathbf{L}^2_{(q)} \simeq 2\,k_F^2 (1 - \cos\theta)/k_F \Rightarrow |\mathbf{L}_{(q)}| \simeq \sqrt{k_F}\,|\theta|$, the decomposition into angular-momentum channels for an $m$-dimensional FS yields a contribution proportional to

$$
\begin{aligned}
t_{ee} & \simeq \frac{e^2\,\Lambda^{x_e}}{2N} \times 2 \int_{\theta > 0} d\theta\, \frac{\theta^{m-1}\,|\mathbf{L}_{(q)}|}{\mathbf{L}^3_{(q)} + \tilde{\alpha}\,|\mathbf{Q}|^{d-m}} = \frac{e^2\,\Lambda^{x_e}}{N\,k_F^{m/2}} \int_{\theta > 0} d|\mathbf{L}_{(q)}|\, \frac{|\mathbf{L}_{(q)}|^m}{\mathbf{L}^3_{(q)} + \tilde{\alpha}\,|\mathbf{Q}|^{d-m}} \\
& = \frac{e^2\,\Lambda^{x_e}\,\Gamma\left(\frac{m+1}{3}\right)\Gamma\left(\frac{2-m}{3}\right)}{3\,N\,k_F^{m/2}\,\tilde{\alpha}^{\frac{2-m}{3}}\,|\mathbf{Q}|^{\frac{(d-m)(2-m)}{3}}} .
\end{aligned}
\tag{3.3}
$$

This result is evidently independent of the specific angular-momentum channel under consideration. For $m = 2 - \delta$ and $d = m + 1 - \gamma\varepsilon$, the expression yields a pole of the form,

$$t_{ee} \simeq \frac{\tilde{e}\, \Lambda^{\gamma\varepsilon + \frac{(2-m)(m-1)}{6}}\, \Gamma\left(\frac{m+1}{3}\right)}{\beta_d^{\frac{2-m}{3}}\, N\, k_F^{m/2}} \frac{1}{\delta}, \tag{3.4}$$

which is equivalent to the logarithmically-divergent term $\frac{e^2 \Lambda^{x_e} \Gamma\left(\frac{m+1}{3}\right)}{N k_F^{m/2}} \ln\left(\frac{\sqrt{k_F}}{\tilde{\alpha}^{\frac{1}{3}} |\mathbf{Q}|^{\frac{d-m}{3}}}\right)$. Upon continuation to the strongly coupled regime at $m = 1$, this term manifests as an infrared (IR) divergence. The corresponding counterterm is given by

$$\begin{aligned} -\frac{\mu^{\lambda_1 \varepsilon}\, \tilde{e}\, v_1}{4\, N\, a\, \epsilon} \sum_{j_1, j_2} \int \left(\prod_{s=1}^{4} dp_s\right) (2\pi)^{d+1}\, \delta^{(d+1)}(p_1 + p_2 - p_3 - p_4) \left(1 - \delta_{j_1, j_2}\right) \\ \times \Big[ \{\bar{\Psi}_{j_1}(p_3) \Psi_{j_2}(p_1)\} \{\bar{\Psi}_{j_1}(p_4) \Psi_{j_2}(p_2)\} \\ - \{\bar{\Psi}_{j_1}(p_3) \sigma_z \Psi_{j_2}(p_1)\} \{\bar{\Psi}_{j_1}(p_4) \sigma_z \Psi_{j_2}(p_2)\} \Big], \\ \lambda_1 = 1 - \frac{a\,(7 - m^2)}{6\,(m+1)}, \quad a\,\epsilon = 2 - m, \quad v_1 = \frac{\Gamma\left(\frac{m+1}{3}\right)}{\beta_d^{\frac{2-m}{3}}}. \end{aligned} \tag{3.5}$$

Using $\mathbf{L}_{(q)}^2 \simeq 2\, k_F^2 (1 - \cos\theta)/k_F \Rightarrow |\mathbf{L}_{(q)}| \simeq \sqrt{k_F}\, |\theta|$, the decomposition into angular momentum channels for an $m$-dimensional FS leads to the contribution being proportional to:

$$\begin{aligned} t_{ee} &\simeq \frac{e^2\, \Lambda^{x_e}}{2\, N} \times 2 \int_{\theta > 0} d\theta\, \frac{\theta^{m-1}\, |\mathbf{L}_{(q)}|}{\mathbf{L}_{(q)}^3 + \tilde{\alpha}\, |\mathbf{Q}|^{d-m}} = \frac{e^2\, \Lambda^{x_e}}{N\, k_F^{m/2}} \int_{\theta > 0} d|\mathbf{L}_{(q)}|\, \frac{|\mathbf{L}_{(q)}|^m}{\mathbf{L}_{(q)}^3 + \tilde{\alpha}\, |\mathbf{Q}|^{d-m}} \\ &= \frac{e^2\, \Lambda^{x_e}\, \Gamma\left(\frac{m+1}{3}\right) \Gamma\left(\frac{2-m}{3}\right)}{3\, N\, k_F^{m/2}\, \tilde{\alpha}^{\frac{2-m}{3}}\, |\mathbf{Q}|^{\frac{(d-m)(2-m)}{3}}}, \end{aligned} \tag{3.6}$$

which is independent of the angular momentum channel. This expression tells us that, for $m = 2 - \delta$ and $d = m + 1 - \gamma\,\varepsilon$, we get a pole (in $\delta$) parametrised as $t_{ee} \simeq \frac{\tilde{e}\, \Lambda^{\gamma\,\varepsilon + (2-m)\,(m-1)} 6}{} \Gamma\left(\frac{m+1}{3}\right) \beta_d^{\frac{m-2}{3}}\, N^{-1}\, k_F^{-m/2}\, \delta^{-1}$, which is equivalent to the logarithmically divergent term, $\frac{e^2\, \Lambda^{x_e}\, \Gamma\left(\frac{m+1}{3}\right)}{N\, k_F^{m/2}} \ln\left(\frac{\sqrt{k_F}}{\tilde{\alpha}^{\frac{1}{3}}\, |\mathbf{Q}|^{\frac{d-m}{3}}}\right)$. This term, when continued to the strongly-coupled regime of $m = 1$, will translate into the infrared divergence there. The resulting counterterm is given by

$$
\begin{aligned}
&\frac{-\mu^{\lambda_1 \varepsilon}\,\tilde{e}\,v_1}{4\,N\,a\,\epsilon} \sum_{j_1,j_2} \int \left( \prod_{s=1}^{4} dp_s \right) (2\pi)^{d+1}\, \delta^{(d+1)}(p_1 + p_2 - p_3 - p_4)\left(1 - \delta_{j_1,j_2}\right) \\
&\qquad \times \Big[ \{\bar{\Psi}_{j_1}(p_3)\Psi_{j_2}(p_1)\}\,\{\bar{\Psi}_{j_1}(p_4)\Psi_{j_2}(p_2)\} \\
&\qquad\quad - \{\bar{\Psi}_{j_1}(p_3)\sigma_z\Psi_{j_2}(p_1)\}\,\{\bar{\Psi}_{j_1}(p_4)\sigma_z\Psi_{j_2}(p_2)\} \Big], \qquad (3.7)
\end{aligned}
$$

where $\lambda_1 = 1 - \frac{a\,(7-m^2)}{6\,(m+1)}$, $a\,\epsilon = 2 - m$, and $v_1 = \Gamma\left(\frac{m+1}{3}\right) \beta_d^{\frac{m-2}{3}}$.

### 3.2.2 One-Loop Diagrams Generating Terms Proportional to $\tilde{e}\,V_S$

The set of one-loop diagrams which can generate terms proportional to $\tilde{e}\,V_S$ are shown in Fig. 3.3. It turns out that only Fig. 3.4a and b contribute [3]. After cancellation with the appropriate diagrams in Fig. 3.4 consisting of counterterm-vertices, their net contribution to the beta-function is captured by $\left(\frac{v_3}{a} + v_4\right) \frac{\tilde{e}\,V_S\,\mu^{(\lambda_2+\gamma)\,\varepsilon}}{N\,\varepsilon}$, where $v_3 = \frac{2\,\gamma_E}{\beta_d^{\frac{2-m}{3}}\,(2\pi)^{\frac{3}{5-m}+m-2}\,\Gamma\left(\frac{m}{2}\right)\Gamma\left(\frac{3}{2(m+1)}\right)\pi}$, $v_4 = \frac{3\,v_3}{m+1} - \frac{2}{3}$, and $\lambda_2 = a\,(m-1)/6$. Furthermore, the contribution from Fig. 3.3a and b, after cancellation with Fig. 3.4a and b, is given by $\left(\frac{v_5}{\gamma} + \frac{v_6}{a}\right) \frac{\tilde{e}\,V_S\,\mu^{\lambda_1 \varepsilon}}{N\,\varepsilon}$, where $v_5 = -\,2^{5-2d+\frac{m}{2}}\,\gamma_E\,\Gamma\left(\frac{m+1}{3}\right) \beta_d^{\frac{m-2}{3}}\,\pi^{-\frac{d}{2}}\,\Gamma^{-1}\left(\frac{d-m}{2}\right)$ and $v_6 = -v_5$. There are two more diagrams in this class, as shown in Fig. 3.5. Due to

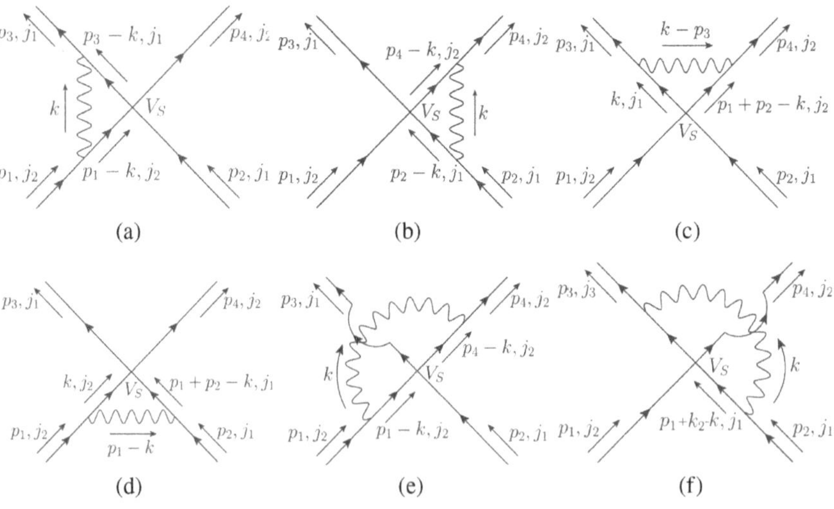


**Fig. 3.3** One-loop diagrams proportional to $\tilde{e}\,V_S$, each consisting of two fermionic and one bosonic propagators forming the loop. Here $p_4 = p_1 + p_2 - p_3$, $\mathbf{p}_1 = \mathbf{p}_3$ and $\mathbf{p}_2 = \mathbf{p}_4$

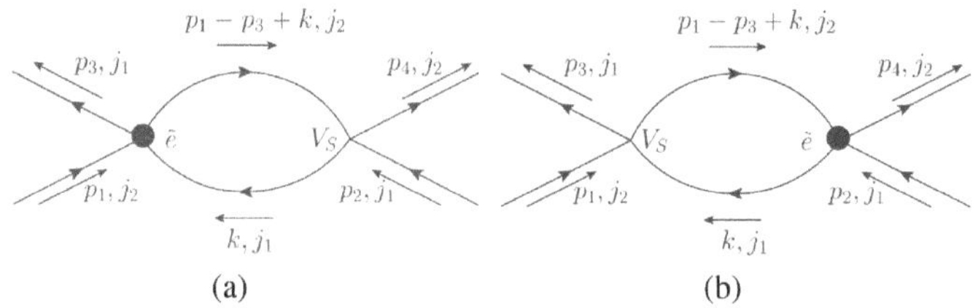


**Fig. 3.4** One-loop diagrams proportional to $\tilde{e}\, V_S$, resulting from counterterms. The counterterm vertex has been denoted by a blob. Here $p_4 = p_1 + p_2 - p_3$

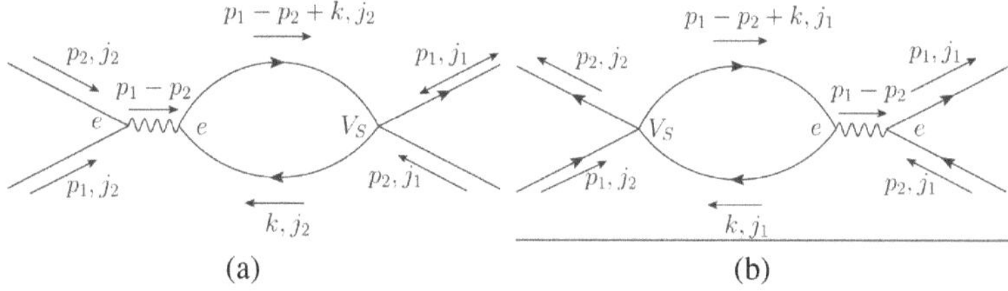


**Fig. 3.5** One-loop diagrams proportional to $e^2\, V_S$, each consisting of two Yukawa vertices and one fermionic loop

the vanishing of the trace of the gamma matrices in the fermionic loop, they are identically zero.

The counterterms essential to account for the four-fermion interactions are captured by

$$S_{CT}^{SC} = -\frac{\mu^{\gamma\,\varepsilon} \tilde{A}_S \tilde{V}_S}{4} \sum_{j_1, j_2} \int \left( \prod_{s=1}^{4} dp_s \right) (2\pi)^{d+1}\, \delta^{(d+1)}(p_1 + p_2 - p_3 - p_4)\left(1 - \delta_{j_1, j_2}\right)$$
$$\times \Big[ \{\bar{\Psi}_{j_1}(p_3) \Psi_{j_2}(p_1)\} \{\bar{\Psi}_{j_1}(p_4) \Psi_{j_2}(p_2)\}$$
$$- \{\bar{\Psi}_{j_1}(p_3) \sigma_z \Psi_{j_2}(p_1)\} \{\bar{\Psi}_{j_1}(p_4) \sigma_z \Psi_{j_2}(p_2)\} \Big],$$

where $\tilde{A}_S \equiv \tilde{Z}_S - 1 = \sum_{\alpha_1=1}^{\infty} \frac{\tilde{Z}_{S,\alpha_1}(\tilde{e}, \tilde{k}_F, V_S)}{\varepsilon^{\alpha_1}}$. On the other hand, denoting the bare quantities by the superscript "$B$", we obtain the bare four-fermion interaction as

$$S_{\text{bare}}^{SC} = -\frac{V_S^{\text{B}}}{4} \sum_{j_1, j_2} \int \left( \prod_{s=1}^{4} dp_s^{\text{B}} \right) (2\pi)^{d+1}\, \delta^{(d+1)}(p_1^{\text{B}} + p_2^{\text{B}} - p_3^{\text{B}} - p_4^{\text{B}})\left(1 - \delta_{j_1, j_2}\right)$$
$$\times \Big[ \{\bar{\Psi}_{j_1}^{\text{B}}(p_3) \Psi_{j_2}^{\text{B}}(p_1)\} \{\bar{\Psi}_{j_1}^{\text{B}}(p_4) \Psi_{j_2}^{\text{B}}(p_2)\}$$
$$- \{\bar{\Psi}_{j_1}^{\text{B}}(p_3) \sigma_z \Psi_{j_2}^{\text{B}}(p_1)\} \{\bar{\Psi}_{j_1}^{\text{B}}(p_4) \sigma_z \Psi_{j_2}^{\text{B}}(p_2)\} \Big], \tag{3.8}$$

where $V_S^{\mathrm{B}}\, k_F^{m/2} = \tilde{Z}_S\, Z_\psi^{-2} \left(\frac{Z_2}{Z_1}\right)^{3(d-m)} \mu^{\gamma\,\varepsilon}\, \tilde{V}_S$. Using the definition $2 - m = a\,\varepsilon \Rightarrow a = \frac{3\,(1-\gamma)}{1+(1-\gamma)\,\varepsilon}$, we have $\frac{\tilde{Z}_{S,1}\, \tilde{V}_S\, \mu^{\gamma\,\varepsilon}}{\varepsilon} = \frac{\left(\frac{v_1}{a} + \frac{v_5 \tilde{V}_S}{\gamma} + \frac{v_6 \tilde{V}_S}{a}\right) \tilde{e}\, \mu^{\lambda_1\,\varepsilon}}{N\,\varepsilon} + \frac{v_2\, \tilde{V}_S^2\, \mu^{\gamma\,\varepsilon}}{\gamma\,\varepsilon} + \frac{\left(\frac{v_3}{a} + v_4\right) \tilde{e}\, \tilde{V}_S\, \mu^{(\lambda_2+\gamma)\varepsilon}}{N\,\varepsilon}$.

Since the bare quantities should not depend on the floating mass scale, $\mu$, we demand that $\frac{d}{d\ln\mu}\left(V_S^{\mathrm{B}}\, k_F^{m/2}\right) = 0$. This leads to the beta-function for $\tilde{V}$ as

$$
\begin{aligned}
&\beta_V + \frac{\left\{\frac{v_1}{a} + \left(\frac{v_3+v_6}{a} + v_4 + \frac{v_5}{\gamma}\right) \tilde{V}_S\right\} \beta_{\tilde{e}} + \left(\frac{v_3+v_6}{a} + v_4 + \frac{v_5}{\gamma}\right) \beta_V\, \tilde{e}}{N\,\varepsilon} \\
&+ \left\{\gamma\,\varepsilon - 4\,\eta_\psi + 3\,(d-m)\,(1-z)\right\} \left(\tilde{V}_S + \frac{v_2\, \tilde{V}_S^2}{\gamma\,\varepsilon}\right) \\
&+ \frac{2\, v_2\, \tilde{V}_S\, \beta_V}{\gamma\,\varepsilon} + \frac{\left\{\lambda_1\,\varepsilon - 4\,\eta_\psi + 3(d-m)(1-z)\right\} \left(\frac{v_1}{a} + \frac{v_5 \tilde{V}_S}{\gamma} + \frac{v_6 \tilde{V}_S}{a}\right) \tilde{e}}{N\,\varepsilon} \\
&+ \left\{(\lambda_2 + \gamma)\,\varepsilon - 4\,\eta_\psi + 3\,(d-m)\,(1-z)\right\} \frac{\left(\frac{v_3}{a} + v_4\right) \tilde{e}\, \tilde{V}_S}{N\,\varepsilon} = 0,
\end{aligned}
\tag{3.9}
$$

Because there are two coupling constants, we have to deal with two beta-functions, $\beta_V$ and $\beta_{\tilde{e}}$. Following the usual procedure, the dependence of the beta-functions on $\epsilon$ must be of the forms, $\beta_V \equiv \frac{\partial \tilde{V}_S}{\partial \ln\mu} = \beta_V^{(0)} + \beta_V^{(1)} \varepsilon$ and $\beta_{\tilde{e}} \equiv \frac{\partial \tilde{e}}{\partial \ln\mu} = -\frac{m+1}{3}\,\varepsilon\,\tilde{e} + \mathcal{O}\left(\tilde{e}^2\right)$. Hence, we need to compare the coefficients of the regular powers of $\varepsilon$, which lead to

$$
\begin{aligned}
\frac{\partial \tilde{V}_S}{\partial l} &= \gamma\,\varepsilon \tilde{V}_S - v_2 \tilde{V}_S^2 - \frac{(7-m^2)\, v_1}{6\,N\,(m+1)}\, \tilde{e} \\
&\quad + \left[\frac{(5 - m^2 - 2m)\,(3\,v_5 - v_6)}{m+1} + (m-1)\, v_3 - 2\,(m+1)\,(u_1 + v_4)\right] \frac{\tilde{e}\, \tilde{V}_S}{6\,N},
\end{aligned}
$$

correct upto $\mathcal{O}\left(\tilde{e}^2, \tilde{e}\,\varepsilon, \varepsilon^2\right)$ at the level one-loop corrections. The final simplified form turns out to be

$$
\frac{\partial \tilde{V}_S}{\partial l} = \begin{cases} \left(\varepsilon - \frac{1}{2}\right) \tilde{V}_S - v_2 \tilde{V}_S^2 - \frac{v_1}{2N}\, \tilde{e} + \frac{3\,v_5 - 4\,v_4 - v_6 - 4\,u_1}{6N}\, \tilde{e}\, \tilde{V}_S & \text{for } m = 1 \\ \gamma\,\varepsilon \tilde{V}_S - v_2 \tilde{V}_S^2 - \frac{v_1}{6N}\, \tilde{e} + \frac{v_3 + 6\,v_4 - 3\,v_5 + v_6 - 6\,u_1}{6N}\, \tilde{e}\, \tilde{V}_S & \text{for } m = 2 \text{ and } \gamma = \pm 1 \end{cases}.
$$

The RG flows for some relevant values of $(d, m)$ have been displayed in Fig. 3.8.

## 3.3 Stability of the Solutions and Final Remarks

The fixed points of the theory are determined by the simultaneous solutions to the flow equations $\frac{\partial \tilde{V}_S}{\partial l} = 0$ and $\frac{\partial \tilde{e}}{\partial l} = 0$. The resulting quadratic equation for the fixed-point values $\tilde{V}_S^*$ yields two distinct roots. Figures 3.6 and 3.7 display the contour

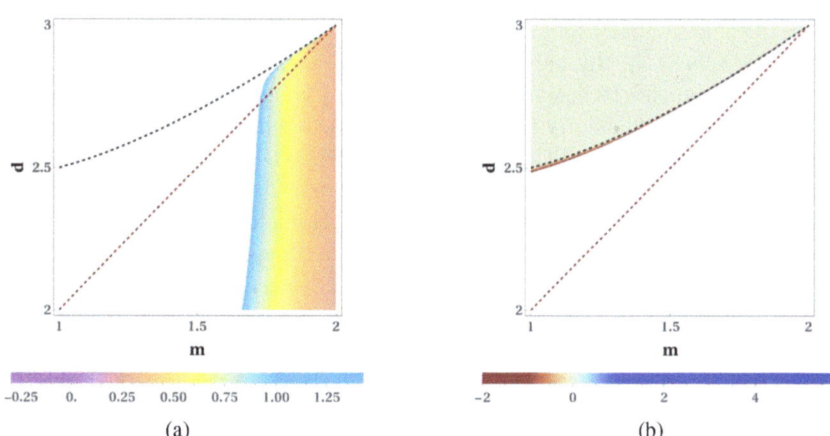


**Fig. 3.6** The contourplots of the two fixed points ($\tilde{V}_S^*$) of $\beta_V$ as functions of $d$ and $m$. The intersection of the white areas in the two density-plots corresponds to regions where no perturbative fixed point exists. The dashed black (red) curve (line) represents $d = d_c (d = \tilde{d}_c)$ in each plot

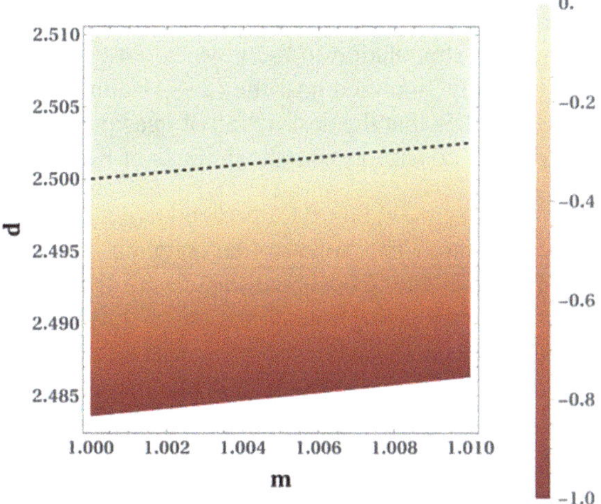


**Fig. 3.7** The contourplot illustrates the second root of the beta-function, $\beta_V = 0$, which corresponds to the fixed-point value $\tilde{V}_S^*$ as a function of the dimensions $m$ and $d$. To highlight the behavior in the vicinity of the physical limit, the region surrounding $m = 1$ has been magnified. The dashed black line denotes the critical dimension, $d = d_c$

plots of these roots as functions of $m$ and $d$, where the analysis is constrained to the perturbative window $|\tilde{V}_S^*| < 1$. This restriction ensures the validity of the underlying expansion. In Fig. 3.6, the intersection of the white regions indicates parameter regimes where no perturbative fixed point exists. In these regimes, the system is invariably driven toward a superconducting state as $\tilde{V}_S^*$ flows into the strong-coupling limit, regardless of the initial bare couplings. To elucidate the stability of these fixed points, representative RG flows for selected values of $m$ and $d$ are presented in Fig. 3.8. Several key observations emerge:

1. At $(d = 3, m = 2)$, the only finite solution is the Gaussian fixed point $(\tilde{V}_S^* = 0, \tilde{e}^* = 0)$, which is IR unstable (cf. Fig. 3.8d). In the absence of gauge fluctuations $(\tilde{e} = 0)$, superconductivity only arises for attractive initial couplings $(\tilde{V}_S < 0)$, consistent with standard BCS theory in a Fermi liquid. However, the introduction of a non-zero coupling $\tilde{e} > 0$ destabilizes the non-Fermi liquid (NFL) phase, driving the system toward strong-coupling superconductivity even for initially repulsive interactions. Consequently, order-parameter fluctuations significantly enhance the superconducting instability relative to the Fermi-liquid benchmark.
2. In the vicinity of $(d = 5/2, m = 1)$ (magnified in Fig. 3.7), a narrow region exists below $d_c$ where two perturbative solutions are present: $(\tilde{V}_S^* = -f_1, \tilde{e}^* = \frac{N\varepsilon}{u_1})$ and the trivial fixed point (see Fig. 3.8a). Here, the NFL phase remains stable for a given initial $\tilde{e} > 0$. Conversely, in the broader vicinity of $(d = 2, m = 1)$, no perturbative fixed points exist, rendering the system unstable to superconductivity for any initial value of $\tilde{V}_S$ (cf. Fig. 3.8c). This behavior holds for $m = 1$ and $d = d_c - \varepsilon$ with $\varepsilon \gtrsim 0.017$.
   The fluctuations of the order parameter thus provide a mechanism for the marked enhancement of pairing in the physical regime of interest near $(d = 2, m = 1)$. While the expansion at $\varepsilon \sim 1/2$ technically lies at the boundary of perturbative control, a continuous extrapolation to the physical case suggests that the pairing instability is strongly amplified near the $(2 + 1)$-dimensional Ising-nematic critical point. This implies that the destruction of quasiparticles is preempted by Cooper pairing, a result consistent with the findings of Ref. [5].

Several remarks regarding the enhanced superconductivity at $(d = 3, m = 2)$ and $(d = 2, m = 1)$ are warranted. The Yukawa-like coupling $\tilde{e}$ generates an effective attractive interaction that mediates Cooper pairing, ensuring that the RG flow targets the strong-coupling superconducting regime for any non-vanishing $\tilde{e}$. The energy scale associated with this instability consistently exceeds the NFL scale at which quasiparticle degradation becomes appreciable. While the magnitude of the pairing gap remains sensitive to the initial couplings, the qualitative behavior aligns with the estimates of Metlitski et al. [5] upon setting $\varepsilon \simeq 0$ and $\varepsilon = 1/2$ for the $(d = 3, m = 2)$ and $(d = 2, m = 1)$ cases, respectively.

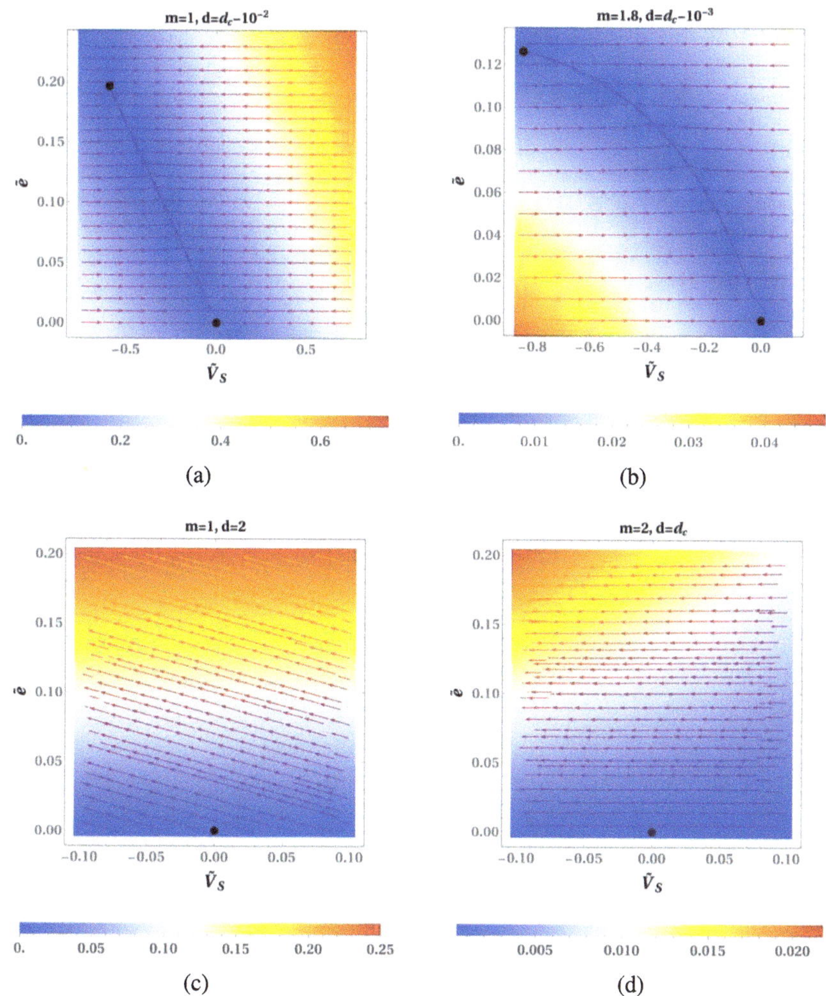


**Fig. 3.8** The representative RG flows in the various regions of the $(d, m)$-plane. The black dots represent the finite fixed points when they exist. The contour-shading conveys the magnitudes of the flow vector $\left(\frac{\partial \tilde{V}_S}{\partial l}, \frac{\partial \tilde{e}}{\partial l}\right)$ in different regions

## References

1. D. Dalidovich, S.S. Lee, Phys. Rev. B **88**, 245106 (2013). https://doi.org/10.1103/PhysRevB.88.245106
2. I. Mandal, S.S. Lee, Phys. Rev. B **92**, 035141 (2015). https://doi.org/10.1103/PhysRevB.92.035141
3. I. Mandal, Phys. Rev. B **94**, 115138 (2016). https://doi.org/10.1103/PhysRevB.94.115138
4. D.T. Son, Phys. Rev. D **59**, 094019 (1999). https://doi.org/10.1103/PhysRevD.59.094019
5. M.A. Metlitski, D.F. Mross, S. Sachdev, T. Senthil, Phys. Rev. B **91**, 115111 (2015). https://doi.org/10.1103/PhysRevB.91.115111

# Chapter 4
# Non-Fermi Liquid Behaviour Induced by Transverse Gauge Fields

## 4.1 Introduction

In this chapter, we consider the case when Fermi surfaces (FSc) are coupled with emergent gauge fields of transverse nature [1–9], leading to non-Fermi-liquid (NFL) behaviour. Like the Ising-nematic quantum critical point (QCP) [10–12], studied in the preceding two chapters, this scenario belongs to the broader family of problems in which the bosonic field's momentum is centred at zero. Nevertheless, the two cases are physically distinct in an important way: At the Ising-nematic transition, the order-parameter boson couples to antipodal patches of the FS with the *same* sign of the coupling strength [13], whereas a transverse gauge field couples to the two antipodal patches with *opposite* signs [10]. This sign difference has profound consequences for the low-energy physics, and motivates a separate treatment. We apply the dimensional regularization procedure to determine the low-energy scaling behavior of an $m$-dimensional FS (with $m \geq 1$) interacting with one or more transverse gauge fields. The analysis is developed first for a single $U(1)$ gauge field and, subsequently, generalized to the $U(1) \times U(1)$ case. The two scenarios are intended to capture two distinct physical situations: a quantum phase transition between a Fermi-liquid (FL) metal and an electrically insulating state devoid of any FS—the deconfined Mott transition—and a transition between two metallic phases whose FSs differ in size on either side of the critical point [14].

## 4.2 Model Involving a Single $U(1)$ Transverse Gauge Field

We first consider an $m$-dimensional FS, which is coupled to a $U(1)$ transverse gauge field $a$ in $d = (m + 1)$ spatial dimensions. The set-up is identical to Ref. [15] discussed in Chap. 1. Using the same patch coordinates, the minimal Euclidean action (in the Matsubara space) is given by [10, 16]



I. Mandal, *Controlled Description Of Non-Fermi Liquids Using Field-Theoretic Methods*, SpringerBriefs in Physics, https://doi.org/10.1007/978-3-032-21618-2_4

$$S = \sum_j \int dk\, \bar{\Psi}_j(k)\, i \left[\mathbf{\Gamma} \cdot \mathbf{K} + \gamma_{d-m}\, \delta_k\right] \Psi_j(k)\, \exp\left\{\frac{\mathbf{L}_{(k)}^2}{\mu\, \tilde{k}_F}\right\}$$
$$+ \frac{1}{2} \int dk\, \mathbf{L}_{(k)}^2\, a^\dagger(k)\, a(k) + \frac{e\, \mu^{\frac{x}{2}}}{\sqrt{N}} \sum_j + \int dk\, dq\, a(q)\, \bar{\Psi}_j(k+q)\, \gamma_0\, \Psi_j(k).$$

Incorporating the one-loop self-energy into the bare bosonic propagator cf. Fig. 4.1a, the dressed propagator takes the form,

$$\Pi_1(k) = -e^2\, \mu^x \int dq\, \mathrm{Tr}\left[\gamma_0\, G_0(k+q)\, \gamma_0\, G_0(q)\right]$$
$$= \frac{-\beta(d,m)\, e^2\, \mu^x \left(\mu\, \tilde{k}_F\right)^{\frac{m-1}{2}} \left[k_0^2 + (m+1-d)\, \tilde{\mathbf{K}}^2\right]}{|\mathbf{L}_{(k)}|\, |\mathbf{K}|^{2-d+m}},$$
$$\text{where } \beta(d,m) = \frac{\pi^{\frac{4-d}{2}}\, \Gamma(d-m)\, \Gamma(m+1-d)}{2^{\frac{4d-m-1}{2}}\, \Gamma^2\left(\frac{d-m+2}{2}\right) \Gamma\left(\frac{m+1-d}{2}\right)}. \tag{4.1}$$

We now turn to the computation of the one-loop fermionic self-energy see Fig. 4.1b, $\Sigma_1(k)$. The final expressions turns out to be

$$\Sigma_1(k) = -\frac{i\, e^{\frac{2(m+1)}{3}} \left[u_0\, \gamma_0\, k_0 + u_1 \left(\tilde{\mathbf{\Gamma}} \cdot \tilde{\mathbf{K}}\right)\right]}{N\, \tilde{k}_F^{\frac{(m-1)(2-m)}{6}}\, \epsilon} + \text{finite terms}, \tag{4.2}$$

where $u_0, u_1 \geq 0$. This diverges logarithmically in $\Lambda$ at the upper critical dimension, $d_c(m) = m + 3/(m+1)$. Within dimensional regularization, this logarithmic divergence in $\Lambda$ manifests as a $1/\epsilon$ pole. For the cases of interest, the numerical values are:

$$\begin{cases} u_0 = 0.0201044, \quad u_1 = 1.85988 & \text{for } m = 1 \\ u_0 = u_1 = 0.0229392 & \text{for } m = 2 \end{cases}. \tag{4.3}$$

The one-loop vertex correction cf. Fig. 4.1c is found to be

$$\Gamma_1(k, 0) = -\frac{e^{\frac{2(m+1)}{3}}\, u_4\, \gamma_0}{N\, \tilde{k}_F^{(m-1)(2-m)/6}\, \epsilon} \left(\frac{\mu}{|\tilde{\mathbf{K}}|}\right)^{\frac{(m+1)\epsilon}{3}} \left[\mathcal{F}\left(\frac{|k_0|}{|\tilde{\mathbf{K}}|}\right)\right]^\epsilon + \text{finite terms}, \tag{4.4}$$

where $u_4 \geq 0$ and $\mathcal{F}$ is a dimensionless function of $|k_0|/|\tilde{\mathbf{K}}|$. The explicit values are

$$u_4 = \begin{cases} 0.0000706373 & \text{for } m = 1 \\ 0 & \text{for } m = 2 \end{cases}. \tag{4.5}$$

This stands in contrast to the Ising-nematic case, where the vertex correction is guaranteed to vanish by a Ward identity [10, 15].

We vary the dimension of the FS from $m = 1$ to $m = 2$ while keeping $\epsilon$ small, thereby obtaining a controlled description for any $m$ in this range. For a given $m$, we tune $d$ so that $\epsilon = d_c(m) - d$ remains small. To absorb the UV divergences arising in the $\epsilon \to 0$ limit, we introduce the following counterterm action:

$$S_{CT} = \sum_j \int dk\, \bar{\Psi}_j(k)\, i\Big[A_0\, \gamma_0\, k_0 + A_1\, \tilde{\boldsymbol{\Gamma}} \cdot \tilde{\mathbf{K}} + A_2\, \gamma_{d-m}\, \delta_k\Big]\Psi_j(k)\, \exp\left\{\frac{\mathbf{L}^2_{(k)}}{\mu\, \tilde{k}_F}\right\}$$
$$+ \frac{A_3}{2}\int dk\, \mathbf{L}^2_{(k)}\, a^\dagger(k)\, a(k) + \frac{A_4\, e\, \mu^{x/2}}{\sqrt{N}} \sum_j \int dk\, dq\, a(q)\, \bar{\Psi}_j(k+q)\, \gamma_0\, \Psi_j(k),$$
$$\text{where}\quad A_\zeta = \sum_{\lambda=1}^{\infty} \frac{Z_\zeta^{(\lambda)}(e, \tilde{k}_F)}{\epsilon^\lambda} \quad \text{with}\quad \zeta = 0, 1, 2, 3, 4. \tag{4.6}$$

The $(d - m - 1)$-dimensional rotational invariance in the space perpendicular to the FS guarantees that each term in $\tilde{\boldsymbol{\Gamma}} \cdot \tilde{\mathbf{K}}$ is renormalized identically, while the sliding symmetry along the FS ensures that the form of $\delta_k$ is preserved under renormalization. The counterterm coefficients $A_0$, $A_1$, and $A_2$ are in general distinct, owing to the absence of full rotational symmetry in the $(d + 1)$-dimensional spacetime. This contrasts with the Ising-nematic case, where $A_0 = A_1$ was enforced by a rotational symmetry acting on the full $(d - m)$-dimensional subspace.

The renormalized action is defined as the physical action free of UV divergences, obtained by subtracting the counterterm action from the bare action, which reads

$$S_{\text{bare}} = \sum_j \int dk^B\, \bar{\Psi}^B_j(k^B)\, i\left[\gamma_0\, k^B_0 + \tilde{\boldsymbol{\Gamma}} \cdot \tilde{\mathbf{K}}^B + \gamma_{d-m}\, \delta_{k^B}\right]\Psi^B_j(k^B)\, \exp\left\{\frac{\mathbf{L}^2_{(k^B)}}{k_{F^B}}\right\}$$
$$+ \frac{1}{2}\int dk^B\, \mathbf{L}^2_{(k^B)}\, a^{B\dagger}(k^B)\, a^B(k^B)$$
$$+ \frac{e^B}{\sqrt{N}} \sum_j \int dk^B\, dq^B\, a^B(q^B)\, \bar{\Psi}^B_j(k^B + q^B)\, \gamma_0\, \Psi^B_j(k^B), \tag{4.7}$$

where

$$k^B_0 = \frac{Z_0}{Z_2}\, k_0, \quad \tilde{\mathbf{K}}^B = \frac{Z_1}{Z_2}\, \tilde{\mathbf{K}}, \quad k^B_{d-m} = k_{d-m}, \quad \mathbf{L}_{(k^B)} = \mathbf{L}_{(k)},$$
$$\Psi^B_j(k^B) = Z_\Psi^{1/2}\, \Psi_j(k), \quad a^B(k^B) = Z_a^{1/2}\, a(k), \quad k^B_F = k_F = \mu\, \tilde{k}_F,$$
$$Z_\Psi = \frac{Z_2^{d-m+1}}{Z_0\, Z_1^{d-m-1}}, \quad Z_a = \frac{Z_3\, Z_2^{d-m}}{Z_0\, Z_1^{d-m-1}}, \quad e^B = Z_e\, e\, \mu^{x/2},$$
$$Z_e = \frac{Z_4\, Z_2^{\frac{d-m}{2}-1}}{\sqrt{Z_0\, Z_3}\, Z_1^{\frac{d-m-1}{2}}}. \tag{4.8}$$

The superscript "$B$" labels bare fields, couplings, and momenta throughout. The bare action in Eq. (4.7) admits a freedom to rescale fields and momenta independently without affecting the physics, and we fix this freedom by requiring $\delta_{k^B} = \delta_k$, which amounts to measuring the scaling dimensions of all other quantities relative to that of $\delta_k$.

The dynamical critical exponent $z$, the critical exponent $\tilde{z}$ along the extra spatial dimensions, the beta-functions $\beta_e$ and $\beta_{k_F}$ for the couplings $e$ and $\tilde{k}_F$, and the anomalous dimensions $\eta_\Psi$ and $\eta_a$ for the fermions and the gauge boson, respectively, are defined by

$$z = 1 + \frac{\partial \ln(Z_0/Z_2)}{\partial \ln \mu}, \quad \tilde{z} = 1 + \frac{\partial \ln(Z_1/Z_2)}{\partial \ln \mu}, \quad \eta_\Psi = \frac{1}{2}\frac{\partial \ln Z_\Psi}{\partial \ln \mu},$$
$$\eta_a = \frac{1}{2}\frac{\partial \ln Z_a}{\partial \ln \mu}, \quad \beta_{k_F}(\tilde{k}_F) = \frac{\partial \tilde{k}_F}{\partial \ln \mu}, \quad \beta_e = \frac{\partial e}{\partial \ln \mu}. \tag{4.9}$$

In the $\epsilon \to 0$ limit, we seek solutions of the form

$$z = z^{(0)}, \quad \tilde{z} = \tilde{z}^{(0)}, \quad \eta_\Psi = \eta_\Psi^{(0)} + \eta_\Psi^{(1)}\,\epsilon, \quad \eta_a = \eta_a^{(0)} + \eta_a^{(1)}\,\epsilon, \quad \beta_e = \beta_e^{(0)} + \beta_e^{(1)}\,\epsilon. \tag{4.10}$$

### 4.2.1 RG Flows at One-Loop Order

Limiting ourselves to one-loop order, the counterterms are parametrised by $Z_\zeta = 1 + \frac{Z_\zeta^{(1)}}{\epsilon}$. Collecting all the results, we find that only $Z_0^{(1)} = -\frac{u_0\,\tilde{e}}{N}$, $Z_1^{(1)} = -\frac{u_1\,\tilde{e}}{N}$, and $Z_4^{(1)} = -\frac{u_4\,\tilde{e}}{N}$ are nonzero, where $\tilde{e} = e^{\frac{2\,(m+1)}{3}}/\tilde{k}_F^{\frac{(m-1)(2-m)}{6}}$. The beta-function for $\tilde{e}$ is given by

$$-\frac{\beta_{\tilde{e}}}{\tilde{e}} = \frac{(m+1)\,\epsilon}{3} - \frac{(m+1)\,[\,(m+1)\,(u_0 - 2\,u_4) + (2-m)\,u_1\,]}{9\,N}\,\tilde{e}. \tag{4.11}$$

The interacting fixed point is obtained from $\beta_{\tilde{e}} = 0$ and $\tilde{e} \neq 0$, leading to

$$\tilde{e}^* = \frac{3\,N\,\epsilon}{(m+1)\,(u_0 - 2\,u_4) + (2-m)\,u_1} + \mathcal{O}\left(\epsilon^2\right). \tag{4.12}$$

It can be checked that this is an IR stable fixed point by computing the first derivative of $\beta_{\tilde{e}}$. The critical exponents at this stable fixed point are given by

$$
\begin{aligned}
z^* &= 1 + \frac{(m+1)\,u_0\,\epsilon}{(m+1)\,(u_0 - 2\,u_4) + (2-m)\,u_1}, \\
\tilde{z}^* &= 1 + \frac{(m+1)\,u_1\,\epsilon}{(m+1)\,(u_0 - 2\,u_4) + (2-m)\,u_1}, \\
\eta_\Psi^* = \eta_a^* &= -\frac{(m+1)\,u_0 + (2-m)\,u_1}{(m+1)\,(u_0 - 2\,u_4) + (2-m)\,u_1}\,\frac{\epsilon}{2}.
\end{aligned}
\tag{4.13}
$$

### *4.2.2 Higher-Loop Corrections*

Let us discuss the implications of the higher-loop corrections, without actually computing the Feynman diagrams. For $m > 1$, we expect a nontrivial UV/IR mixing to be present, as was found in Ref. [11, 15], which makes the results one-loop exact. In other words, all higher-loop corrections would vanish for $m > 1$ in the limit $k_F \to 0$, due to suppression of the results by positive powers of $k_F$. For $m = 1$, we will use the arguments and results of Ref. [10] to assume a generic form of the corrections coming from two-loop diagrams. Henceforth, we will just focus on $m = 1$ in this subsection.

The two-loop diagrams for the boson self-energy turn out to be UV finite and hence, renormalize, the factor $\beta(\frac{5}{2}, 1)$ by a finite amount $\beta_2 = \kappa\,\tilde{e}/N$, where $\kappa$ is a finite number. Consequently, the bosonic propagator at this order will takes the form of $D_2(q) = \left[ \mathbf{L}_{(q)}^2 + \frac{\left[\beta\left(\frac{5}{2},2\right) + \frac{\kappa\,\tilde{e}}{N}\right] e^2 \mu^\epsilon}{|\mathbf{L}_{(q)}|} \times \frac{k_0^2 + \left(\epsilon - \frac{1}{2}\right) \tilde{\mathbf{K}}^2}{|\mathbf{K}|^{\frac{1}{2}+\epsilon}} \right]^{-1}$. From this, the fermion self-energy now receives a correction of $\Sigma_2^{(1)}(k) = \left[ \left\{ \frac{\beta\left(\frac{5}{2},2\right)}{\beta\left(\frac{5}{2},2\right) + \frac{\kappa\,\tilde{e}}{N}} \right\}^{1/3} - 1 \right] \Sigma_1(k) =$ $-\,\kappa\,\tilde{e}\,\Sigma_1(k)/[3\,N\,\beta(\frac{5}{2}, 2)] +$ finite terms. Now the two-loop fermion self-energy diagrams, after taking into account the counterterms obtained from one-loop corrections, take the form of $\Sigma_2^{(2)}(k) = -\,i\,\tilde{e}^2 \left[ \tilde{v}_0\,\gamma_0\,q_0 + \tilde{v}_1\,\left(\tilde{\mathbf{\Gamma}} \cdot \tilde{\mathbf{Q}}\right) + w\,\gamma_{d-1}\,\delta_k \right] / (N^2 \epsilon)$ + finite terms. Adding the two parts, the generic form of the total two-loop fermion self-energy can be written as $\Sigma_2^{tot}(k) = -i\tilde{e}^2 \left[ v_0\,\gamma_0\,q_0 + v_1 \tilde{\mathbf{\Gamma}} \cdot \tilde{\mathbf{Q}} + w\,\gamma_{d-1}\,\delta_k \right] /$ $(N^2 \epsilon)$ + finite terms, where $v_0 = u_0 + \tilde{v}_0$ and $v_1 = u_1 + \tilde{v}_1$.

There will also be a divergent vertex correction (see Fig. 4.1c) which will lead to a nonzero $Z_4^{(1)}$ of the form $-\frac{\tilde{e}^2\,y}{N^2}$. All these now lead to the nonzero coefficients,

$$
\begin{aligned}
Z_0^{(1)} &= -\frac{u_0\,\tilde{e}}{N} - \frac{v_0\,\tilde{e}^2}{N^2}, \quad Z_1^{(1)} = -\frac{u_1\,\tilde{e}}{N} - \frac{v_1\,\tilde{e}^2}{N^2}, \\
Z_2^{(1)} &= -\frac{w\,\tilde{e}^2}{N^2}, \quad Z_4^{(1)} = -\frac{u_4\,\tilde{e}}{N} - \frac{y\,\tilde{e}^2}{N^2},
\end{aligned}
\tag{4.14}
$$

resulting in

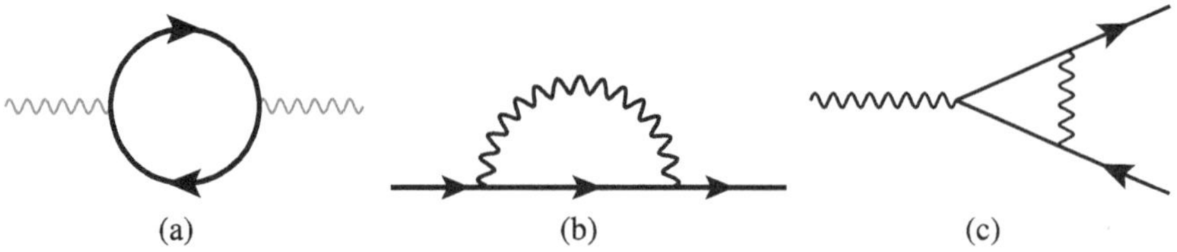


**Fig. 4.1** One-loop diagrams for the **a** bosonic self-energy, **b** fermionic self-energy, and **c** vertex correction. In **a**, the bare bosonic propagator is represented by a gray wiggly curve. In **b** and **c**, solid arrowed lines denote bare fermionic propagators, while the wiggly curves depict dressed bosonic propagators that incorporate the one-loop self-energy shown in (**a**)

$$\begin{aligned}\frac{\beta_{\tilde{e}}}{\tilde{e}} &= -\frac{2\,(2\,u_1\tilde{e} + 3\,N)\,\epsilon}{9\,N} - \frac{2\,(2\,u_0 + u_1 - 4\,u_4)\,\tilde{e}}{9\,N} \\ &+ \frac{4\left[-u_1^2 - 2\,u_0\,u_1 + 4\,u_4\,u_1 - 3\,(2\,v_0 + v_1 - 3\,w) + 12\,y\right]\tilde{e}^2}{27\,N^2}.\end{aligned} \tag{4.15}$$

At the fixed point, we now have

$$\frac{\tilde{e}^*}{N} = \frac{3\,\epsilon}{2\,u_0 + u_1 - 4\,u_4} - \frac{18\,(2\,v_0 + v_1 - 3\,w - 4\,y)\,\epsilon^2}{(2\,u_0 + u_1 - 4\,u_4)^3} + \mathcal{O}(\epsilon^3). \tag{4.16}$$

This shows that the nature of the stable non-Fermi liquid fixed point remains unchanged, although its location (as well as any critical scaling) gets corrected by one higher power of $\epsilon$.

### 4.2.3 Renormalization of the $2k_F$ Scattering Amplitude

In order to examine how the back-scattering is affected by the interactions with the gauge bosons in the non-Fermi liquid state, we consider an operator which carries momentum $2k_F$ as follows:

$$\begin{aligned}S_{2k_F} &= -2\,g_{2k_F}\,\mu \sum_j \int dk \left[(\psi^\dagger_{+,j}(k)\psi_{-,j}(k) + \psi^\dagger_{-,j}(k)\psi_{+,j}(k)\right] \\ &= i\,g_{2k_F}\,\mu \int dk \left[\Psi^T(k)\gamma_0\Psi(-k) + \bar{\Psi}(k)\gamma_0\bar{\Psi}^T(-k)\right],\end{aligned} \tag{4.17}$$

where $g_{2k_F}$ is the source. To cancel UV divergences, we need to add a counterterm of the form

$$S^{CT}_{2k_F} = i\,g_{2k_F}\,\mu\left(Z_{2k_F} - 1\right) \int dk \Big[\Psi^T(k)\,\gamma_0\Psi(-k) + \bar{\Psi}(k)\,\gamma_0\bar{\Psi}^T(-k)\Big], \tag{4.18}$$

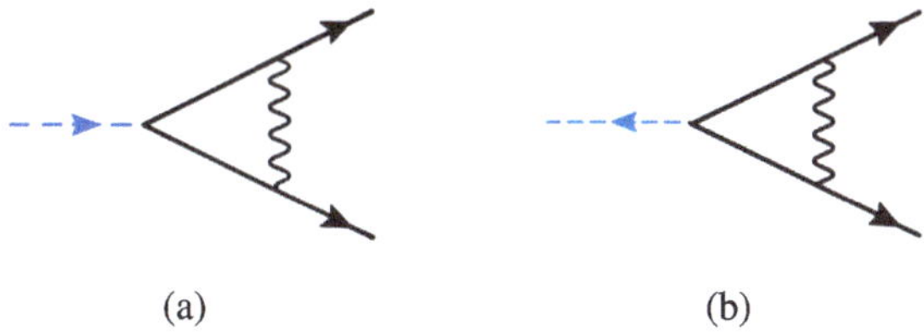

**Fig. 4.2** The one-loop diagrams contributing to the $2k_F$ scattering amplitude

which renormalizes the insertion starting from the bare one,

$$S^{bare}_{2k_F} = i\, g^B_{2k_F} \int dk^B \Big[ \left(\Psi^B(k)\right)^T \gamma_0\, \Psi^B(-k) + \bar{\Psi}^B(k)\, \gamma_0 \left(\bar{\Psi}^B(-k)\right)^T \Big].$$

Here, $g^B_{2k_F} = Z_g\, g_{2k_F}$, $Z_{2k_F} = Z_g\, Z_2$, and $Z_{2k_F} = 1 + Z^{(1)}_{2k_F}/\epsilon$ to one-loop order. The loop calculations, involving the diagrams as shown in Fig. 4.2, lead to

$$Z^{(1)}_{2k_F} = \begin{cases} -\frac{0.0774559\,\tilde{e}}{N} & \text{for } m = 1 \\ 0 & \text{for } m = 2 \end{cases}. \tag{4.19}$$

This gives the beta-function for $g_{2k_F}$ as

$$\beta_g = -g_{2k_F}\left(1 - \eta_g\right), \tag{4.20}$$

with anomalous dimension $\eta_g = -\frac{2\,\tilde{e}\,u_g}{3N}$, where $u_g = 0.0774559$ for $m = 1$. The negative value of $\eta_g$ shows that the $2k_F$ scattering amplitude is enhanced by fluctuations of the transverse gauge field. This is in contrast with the behaviour computed in the case of in the Ising-nematic quantum criticality, where the $2\,k_F$ scattering amplitude is suppressed [10] in the presence of the Ising-nematic critical bosons in $d_p = 2$. Note that this scattering amplitude can also be interpreted as an instability in the charge density wave (CDW) channel, which therefore (due to its negative anomalous dimension) turns out to be a serious competitor for the transverse gauge field criticality for $m = 1$.

### *4.2.4 Thermodynamic Quantities*

The scaling of thermodynamic quantities are different from observables which are local in momentum space. This is because all low energy modes near the FS contribute to the thermodynamic responses. Here, we will outline the expectations of a general scaling analysis. For the critical FS, the momentum components, $k_{d-m}$ and $\mathbf{L}_{(k)}$, each has scaling-dimension one, $k_0$ has scaling dimension $z$, and the remaining momentum components with linear dispersion have scaling dimension $\tilde{z}$. Note that for the Ising-nematic critical point [10, 15], $\tilde{z} = z$.

In order to examine the scaling behavior of thermodynamic quantities, we consider the free energy density at finite temperature $T$. In a system with $d_p = m + 1$ spatial dimensions, fermionic dynamical critical exponent $z$, and $\frac{2-m}{m+1} - \epsilon$ auxiliary dimensions with critical exponent $\tilde{z}$, the free energy density $F(T)$ has the scaling dimension $[F(T)] = d_p + z + \left(\frac{2-m}{m+1} - \epsilon\right)\tilde{z}$, if it were independent of any UV cut-off scale. However, when a critical boson couples with fermions on all parts of the FS, the entire FS becomes hot. As a result, we expect a hyperscaling violation, such that the singular part of the free energy density depends on the size of the FS [17, 18]. The largest momentum along the $\mathbf{L}_{(k)}$ direction is set by the Fermi momentum $k_F$, and hence the free energy density should have the following scaling form:

$$F(T) \sim k_F^{m/2}\, T^{1+\frac{d_p-m}{z}+\frac{\left(\frac{2-m}{m+1}-\epsilon\right)\tilde{z}}{z}} \sim k_F^{m/2}\, T^{1+\frac{1+\left(\frac{2-m}{m+1}-\epsilon\right)\tilde{z}}{z}} \tag{4.21}$$

in the presence of an $m$-dimensional FS, with an effective scaling dimension, $[F(T)]_{\text{eff}} = 1 + z + \left(\frac{2-m}{m+1} - \epsilon\right)\tilde{z}$. From this scaling form, we can extract the temperature dependence of various observables within the quantum critical region. For example, the specific heat should scale as $C \propto T^{\frac{1+\left(\frac{2-m}{m+1}-\epsilon\right)\tilde{z}}{z}}$.

The current operator is given by $J(T) = \frac{\delta F(T)}{\delta A}$, where $A$ is the vector potential with scaling dimension one. Hence, it should have the scaling form:

$$J(T) \sim k_F^{m/2}\, T^{1+\frac{\left(\frac{2-m}{m+1}-\epsilon\right)\tilde{z}}{z}}, \tag{4.22}$$

with an effective scaling dimension $[J(T)]_{\text{eff}} = z + \left(\frac{2-m}{m+1} - \epsilon\right)\tilde{z}$. Then using the Kubo formula, we can infer that the effective scaling dimension of the optical conductivity is

$$\begin{aligned} [\sigma(\omega)]_{\text{eff}} &= 2\,[J(T)]_{\text{eff}} - z - [\text{volume in } k\text{-space}]_{\text{eff}} \\ &= 2z + 2\left(\frac{2-m}{m+1} - \epsilon\right)\tilde{z} - z - z - 1 - \left(\frac{2-m}{m+1} - \epsilon\right)\tilde{z} \\ &= -1 + \left(\frac{2-m}{m+1} - \epsilon\right)\tilde{z}, \end{aligned} \tag{4.23}$$

leading to the scaling form:

$$[\sigma(\omega \gg T)] \propto \omega^{-\frac{1}{z}+\frac{\left(\frac{2-m}{m+1}-\epsilon\right)\tilde{z}}{z}}, \tag{4.24}$$

where $\omega$ is the frequency of the applied AC electric field.

## 4.3 Model Involving Two $U(1)$ Transverse Gauge Fields

In this section, we consider the $m$-dimensional FSs of two different kinds of fermions (denoted by subscripts 1 and 2) coupled to two U(1) gauge fields, $a_c$ and $a_s$, in the context of deconfined Mott transition and deconfined metal-metal transition studied in Ref. [14] (for $m = 1$). The theoretical motivation of Ref. [14] was to study a distinct class of quantum phase transitions between a Fermi liquid and a Mott insulator [19], or between two metals that have FSs with finite but different sizes on either side of the transition [20, 21]. These have been dubbed by the authors as deconfined Mott transition (DMT), and deconfined metal-metal transition (DM$^2$T), respectively. These problems can be formulated using a fictitious/emergent $U(2)$ gauge field, but the authors showed that this non-abelian gauge field is 'quasi-abelianized' such that a related $U(1) \times U(1)$ gauge theory can capture many essential features. In this $U(1) \times U(1)$ gauge theory, the fermion fields $\psi_{1,\pm,j}$ and $\psi_{2,\pm,j}$ carry negative charges under the even $(a_c + a_s)$ and odd $(a_c - a_s)$ combinations of the gauge fields. We revisit this problem using our dimensional regularization scheme because using this technique, we can study this system in generic dimensions, and also perform higher-loop diagrams giving order by order corrections in $\epsilon$.

The action takes the form of

$$
\begin{aligned}
S = & \sum_{\alpha=1,2} \sum_{p=\pm} \sum_{j=1}^{N} \int dk\, \psi^{\dagger}_{\alpha,p,j}(k) \Big[ i\, k_0 + p\, k_{d-m} + \mathbf{L}^2_{(k)} \Big] \psi_{\alpha,p,j}(k) \\
& + \frac{1}{2} \int dk\, \mathbf{L}^2_{(k)} \left[ a^{\dagger}_c(k)\, a_c(k) + a^{\dagger}_s(k)\, a_s(k) \right] \\
& + \sum_{\alpha=1,2} \sum_{p=\pm} p \sum_{j=1}^{N} \int dk\, dq \Big[ \frac{(-1)^{\alpha}\, e_s}{\sqrt{N}}\, a_s(q)\, \psi^{\dagger}_{\alpha,p,j}(k+q)\, \psi_{\alpha,p,j}(k) \\
& \qquad - \frac{e_c}{\sqrt{N}}\, a_c(q)\, \psi^{\dagger}_{\alpha,p,j}(k+q)\, \psi_{\alpha,p,j}(k) \Big], \qquad (4.25)
\end{aligned}
$$

where $e_c$ and $e_s$ denote the gauge couplings for the gauge fields $a_c$ and $a_s$ respectively. We will perform dimensional regularization on this action and determine the RG fixed points. Our formalism allows us to extend the discussion beyond $m = 1$, and also to easily compute higher-loop corrections.

### *4.3.1 Dimensional Regularization*

Proceeding as in the single transverse gauge field case, we add artificial co-dimensions for dimensional regularization after introducing the two-component spinors,

$$\Psi^T_{\alpha,j}(k) = \left(\psi_{\alpha,+,j}(k), \psi^\dagger_{\alpha,-,j}(-k)\right) \text{ and } \bar{\Psi}_{\alpha,j} \equiv \Psi^\dagger_{\alpha,j}\,\gamma_0, \text{ with } \alpha = 1, 2. \quad (4.26)$$

The dressed gauge boson propagators include the one-loop self-energies given by:

$$\Pi^c_1(k) = -\frac{\beta(d,m)\,e_c^2\,\mu^x\left(\mu\,\tilde{k}_F\right)^{\frac{m-1}{2}}}{|\mathbf{L}_{(q)}|}\left[k_0^2 + (m+1-d)\,\tilde{\mathbf{K}}^2\right]|\mathbf{K}|^{d-m-2} \text{ and}$$
$$\Pi^s_1(k) = -\frac{\beta(d,m)\,e_s^2\,\mu^x\left(\mu\,\tilde{k}_F\right)^{\frac{m-1}{2}}}{|\mathbf{L}_{(q)}|}\left[k_0^2 + (m+1-d)\,\tilde{\mathbf{K}}^2\right]|\mathbf{K}|^{d-m-2}, \quad (4.27)$$

for the $a_c$ and $a_s$ gauge fields, respectively. This implies that the one-loop fermion self-energy for both $\Psi_{1,j}$ and $\Psi_{2,j}$ now takes the form,

$$\Sigma_1(q) = -\frac{i\left(e_c^{\frac{2(m+1)}{3}} + e_s^{\frac{2(m+1)}{3}}\right)}{N\,\tilde{k}_F^{\frac{(m-1)(2-m)}{6}}}\,\frac{u_0\,\gamma_0\,q_0 + u_1\left(\tilde{\mathbf{\Gamma}}\cdot\tilde{\mathbf{Q}}\right)}{\epsilon} + \text{finite terms}, \quad (4.28)$$

with the critical dimension $d_c = \left(m + \frac{3}{m+1}\right)$, $u_0$ and $u_1$ having the same values as for the $U(1)$ case.

The counterterms take the same form as the original local action, viz.

$$\begin{aligned}
S_{CT} =& \sum_{\alpha,j}\int dk\,\bar{\Psi}_{\alpha,j}(k)\,i\left[A_0\,\gamma_0\,k_0 + A_1\,\tilde{\mathbf{\Gamma}}\cdot\tilde{\mathbf{K}} + A_2\,\gamma_{d-m}\,\delta_k\right]\Psi_{\alpha,j}(k)\,\exp\left\{\frac{\mathbf{L}^2_{(k)}}{\mu\,\tilde{k}_F}\right\}\\
&+ \frac{A_{3_s}}{2}\int dk\,\mathbf{L}^2_{(k)}\,a_s^\dagger(k)\,a_s(k) + \frac{A_{3_c}}{2}\int dk\,\mathbf{L}^2_{(k)}\,a_c^\dagger(k)\,a_c(k)\\
&- A_{4_c}\frac{e_c\,\mu^{x/2}}{\sqrt{N}}\sum_{\alpha,j}\int dk\,dq\,a_c(q)\,\bar{\Psi}_{\alpha,j}(k+q)\,\gamma_0\,\Psi_{\alpha,j}(k)\\
&+ A_{4_s}\frac{e_s\,\mu^{x/2}}{\sqrt{N}}\sum_{\alpha,j}(-1)^\alpha\int\frac{d^{d+1}k\,d^{d+1}q}{(2\pi)^{2d+2}}\,a_s(q)\,\bar{\Psi}_{\alpha,j}(k+q)\,\gamma_0\,\Psi_{\alpha,j}(k)\,,
\end{aligned} \quad (4.29)$$

where

$$A_\zeta = \sum_{\lambda=1}^{\infty}\frac{Z_\zeta^{(\lambda)}(e,\tilde{k}_F)}{\epsilon^\lambda} \text{ with } \zeta = 0, 1, 2, 3_c, 3_s, 4_c, 4_s. \quad (4.30)$$

We have taken into account the exchange symmetry: $\Psi_{1,j} \leftrightarrow \Psi_{2,j},\ a_s \to -a_s$, which was assumed in Ref. [14], and here it means that both $\Psi_{1,j}$ and $\Psi_{2,j}$ have the same wavefunction renormalization $Z_{\Psi}^{1/2}$. To obtain the renormalized action, not containing any divergences, we subtract the counterterms from the bare action,

$$
\begin{aligned}
S_{bare} = & \sum_{\alpha,j} \int dk^B \, \bar{\Psi}^B_{\alpha,j}(k^B) \, i \left[ \gamma_0 \, k_0^B + \tilde{\mathbf{\Gamma}} \cdot \tilde{\mathbf{K}}^B + \gamma_{d-m} \, \delta_k \right] \Psi^B_{\alpha,j}(k^B) \, \exp\left\{ \frac{\mathbf{L}^2_{(k^B)}}{\mu \, \tilde{k}_F^B} \right\} \\
& + \frac{1}{2} \int dk^B \, \mathbf{L}^2_{(k^B)} \, a_c^{B\dagger}(k^B) \, a_c^B(k^B) + \frac{1}{2} \int dk^B \, \mathbf{L}^2_{(k^B)} \, a_s^{B\dagger}(k^B) \, a_s^B(k^B) \\
& - \frac{e_c^B}{\sqrt{N}} \sum_{\alpha,j} \int dk^B \, dq^B \, a_c^B(q^B) \, \bar{\Psi}^B_{\alpha,j}(k^B + q^B) \, \gamma_0 \, \Psi^B_{\alpha,j}(k^B) \\
& + \frac{e_s^B}{\sqrt{N}} \sum_{\alpha,j} (-1)^{\alpha} \int dk^B \, dq^B \, a_s^B(q^B) \, \bar{\Psi}^B_{\alpha,j}(k^B + q^B) \, \gamma_0 \, \Psi^B_{\alpha,j}(k^B) \,,
\end{aligned}
\tag{4.31}
$$

remembering that $\delta_{k^B} = \delta_k$. Here,

$$
\begin{aligned}
& k_0^B = \frac{Z_0}{Z_2} \, k_0, \quad \tilde{\mathbf{K}}^B = \frac{Z_1}{Z_2} \, \tilde{\mathbf{K}}, \quad k_{d-m}^B = k_{d-m}, \quad \mathbf{L}_{(k^B)} = \mathbf{L}_{(k)}, \quad k_F^B = k_F = \mu \, \tilde{k}_F, \\
& \Psi_j^B(k^B) = Z_{\Psi}^{\frac{1}{2}} \, \Psi_j(k), \quad a_c^B(k^B) = Z_{a_c}^{\frac{1}{2}} \, a_c(k), \quad a_s^B(k^B) = Z_{a_s}^{\frac{1}{2}} \, a_s(k), \\
& Z_{\Psi} = \frac{Z_2^{d-m+1}}{Z_0 \, Z_1^{d-m-1}}, \quad Z_{a_c} = \frac{Z_{3_s} \, Z_2^{d-m}}{Z_0 \, Z_1^{d-m-1}}, \quad Z_{a_s} = \frac{Z_{3_c} \, Z_2^{d-m}}{Z_0 \, Z_1^{d-m-1}}, \quad e_c^B = Z_{e_c} \, e_c \, \mu^{\frac{x}{2}}, \\
& Z_{e_c} = \frac{Z_4 \, Z_2^{\frac{d-m}{2}-1}}{\sqrt{Z_0 \, Z_{3_c}} \, Z_1^{\frac{d-m-1}{2}}}, \quad e_s^B = Z_{e_s} \, e_s \, \mu^{\frac{x}{2}}, \quad Z_{e_s} = \frac{Z_4 \, Z_2^{\frac{d-m}{2}-1}}{\sqrt{Z_0 \, Z_{3_s}} \, Z_1^{\frac{d-m-1}{2}}},
\end{aligned}
\tag{4.32}
$$

and $Z_{\varsigma} = 1 + A_{\varsigma}$. As before, the superscript "B" denotes the bare fields, couplings, and momenta.

As before, we will use the same notations, namely, $z$ for the dynamical critical exponent, $\tilde{z}$ for the critical exponent along the extra spatial dimensions, $\beta_{k_F}$ for the beta-function for $\tilde{k}_F$, and $\eta_{\psi}$ for the anomalous dimension of the fermions. Since we have two gauge fields now, we will use the symbols $\beta_{e_c}$ and $\beta_{e_s}$ to denote the beta-functions for the couplings $e_c$ and $e_s$ respectively, which are explicitly given by

$$
\beta_{e_c} = \frac{\partial e_c}{\partial \ln \mu}, \quad \beta_{e_s} = \frac{\partial e_s}{\partial \ln \mu}. \tag{4.33}
$$

The anomalous dimensions of these two bosons are indicated by

$$
\eta_{a_c} = \frac{1}{2} \frac{\partial \ln Z_{a_c}}{\partial \ln \mu}, \quad \eta_{a_s} = \frac{1}{2} \frac{\partial \ln Z_{a_s}}{\partial \ln \mu}. \tag{4.34}
$$

### 4.3.2 RG Flows at One-Loop Order

At one-loop order, the counterterms are given by $Z_\zeta = 1 + Z_\zeta^{(1)}/\epsilon$, where the non-vanishing coefficients are

$$Z_0^{(1)} = -\frac{u_0\,(\tilde{e}_c + \tilde{e}_s)}{N}, \quad Z_1^{(1)} = -\frac{u_1\,(\tilde{e}_c + \tilde{e}_s)}{N}, \quad Z_4^{(1)} = -\frac{u_4\,(\tilde{e}_c + \tilde{e}_s)}{N}, \tag{4.35}$$

with the effective couplings defined as

$$\tilde{e}_c = \frac{e_c^{\frac{2(m+1)}{3}}}{\tilde{k}_F^{\frac{(m-1)(2-m)}{6}}} \quad \text{and} \quad \tilde{e}_s = \frac{e_s^{\frac{2(m+1)}{3}}}{\tilde{k}_F^{\frac{(m-1)(2-m)}{6}}}. \tag{4.36}$$

The one-loop beta-functions are given by

$$\begin{aligned}
&\beta_{k_F} = -\tilde{k}_F, \quad (1-z)\,Z_0 = -\beta_{e_c}\,\frac{\partial Z_0}{\partial e_c} - \beta_{e_s}\,\frac{\partial Z_0}{\partial e_s} + \tilde{k}_F\,\frac{\partial Z_0}{\partial \tilde{k}_F}, \\
&(1-\tilde{z})\,Z_1 = -\beta_{e_c}\,\frac{\partial Z_1}{\partial e_c} - \beta_{e_s}\,\frac{\partial Z_1}{\partial e_s} + \tilde{k}_F\,\frac{\partial Z_1}{\partial \tilde{k}_F}, \quad \frac{\beta_{e_c}}{e_c} = -\frac{\epsilon}{2} + \frac{1}{2}\left[\frac{(2-m)\tilde{z}}{m+1} + z - 2 + \frac{m}{2}\right], \\
&\frac{\beta_{e_s}}{e_s} = -\frac{\epsilon}{2} + \frac{1}{2}\left[\frac{(2-m)\tilde{z}}{m+1} + z - 2 + \frac{m}{2}\right].
\end{aligned} \tag{4.37}$$

Solving these equations yields

$$\begin{aligned}
-\frac{\beta_{e_c}}{e_c} = -\frac{\beta_{e_s}}{e_s} = \frac{\epsilon}{2} &+ \frac{(m-1)(2-m)}{4(m+1)} \\
&- \frac{(m+1)\,u_0 + (2-m)\,u_1 - 2(m+1)\,u_4}{6N}\,(\tilde{e}_c + \tilde{e}_s)\,.
\end{aligned} \tag{4.38}$$

As in the single gauge field case, the order-by-order loop corrections for generic $m$ are controlled not by the bare couplings $e_c$ and $e_s$ themselves, but by the effective couplings $\tilde{e}_c$ and $\tilde{e}_s$. The RG flows are therefore most naturally expressed through the beta-functions of these effective couplings as follows:

$$-\frac{\beta_{\tilde{e}_c}}{\tilde{e}_c} = -\frac{\beta_{\tilde{e}_s}}{\tilde{e}_s} = \frac{(m+1)\,\epsilon}{3} - \frac{(m+1)\,[(m+1)(u_0 - 2u_4) + (2-m)u_1]\,(\tilde{e}_c + \tilde{e}_s)}{9N}. \tag{4.39}$$

The interacting fixed points are located at the zeros of these beta-functions, and satisfy

$$\tilde{e}_c^* + \tilde{e}_s^* = \frac{3N\,\epsilon}{(m+1)(u_0 - 2u_4) + (2-m)u_1} + \mathcal{O}(\epsilon^2), \tag{4.40}$$

which defines a fixed *line* in the space of couplings, consistent with the finding of Ref. [14] for the special case $m = 1$. That this fixed line is infrared stable can be verified by examining the first derivatives of the beta-functions. We have thus established that the fixed line feature persists for critical FSs of dimension greater than one. The critical exponents at this stable fixed line take the same form as those in Eq. (4.13).

### 4.3.3 Corrections Arising from Feynman Diagrams with More Than One Loop

Applying the same reasoning as in the single gauge field case, the nonvanishing $Z_\zeta^{(1)}$ coefficients, including one- and two-loop corrections for $m = 1$, are found to be

$$
\begin{aligned}
Z_0^{(1)} &= -\frac{u_0\,(\tilde{e}_c+\tilde{e}_s)}{N} - \frac{v_0\,(\tilde{e}_s+\tilde{e}_c)^2}{N^2}, \quad Z_1^{(1)} = -\frac{u_1\,(\tilde{e}_c+\tilde{e}_s)}{N} - \frac{v_1\,(\tilde{e}_c+\tilde{e}_s)^2}{N^2},\\
Z_2^{(1)} &= -\frac{w\,(\tilde{e}_c+\tilde{e}_s)^2}{N^2}, \quad Z_{4_s}^{(1)} = -\frac{u_4\,(\tilde{e}_c+\tilde{e}_s)}{N} - \frac{y\,(\tilde{e}_c+\tilde{e}_s)^2}{N^2},\\
Z_{4_c}^{(1)} &= -\frac{u_4\,(\tilde{e}_c+\tilde{e}_s)}{N} - \frac{y\,(\tilde{e}_c+\tilde{e}_s)^2}{N^2}.
\end{aligned}
\tag{4.41}
$$

These lead to the beta-functions

$$
\begin{aligned}
-\frac{\beta_{\tilde{e}_c}}{\tilde{e}_c} = -\frac{\beta_{\tilde{e}_s}}{\tilde{e}_s} = {}& \frac{2\left[2u_1\,(\tilde{e}_c+\tilde{e}_s)+3N\right]\epsilon}{9N} + \frac{2\,(2u_0+u_1-4u_4)\,(\tilde{e}_c+\tilde{e}_s)}{9N}\\
&- \frac{4\left[-u_1^2-2u_0u_1+4u_4u_1-3(2v_0+v_1-3w)+12y\right](\tilde{e}_c+\tilde{e}_s)^2}{27N^2},
\end{aligned}
\tag{4.42}
$$

which again admit a continuous line of fixed points, defined by

$$
\frac{\tilde{e}_s^*+\tilde{e}_c^*}{N} = \frac{3\,\epsilon}{2u_0+u_1-4u_4} - \frac{18\,(2v_0+v_1-3w-4y)\,\epsilon^2}{(2u_0+u_1-4u_4)^3} + \mathcal{O}(\epsilon^3). \tag{4.43}
$$

At three-loop order, however, new contributions to the beta-functions arise that are not simply proportional to integer powers of $(\tilde{e}_s+\tilde{e}_c)$. Representative examples are the Aslamazov-Larkin-type diagrams. The ones contributing to the self-energy of the $a_c$ transverse gauge field in the particle-particle channel are shown in Fig. 4.3. All sixteen such diagrams together yield a contribution proportional to $4\,e_c^2(\tilde{e}_c^2+\tilde{e}_s^2)$. Analogously, the Aslamazov-Larkin diagrams in the particle-hole channel contribute a term proportional to $4\,e_c^2(\tilde{e}_c^2+\tilde{e}_s^2)$, and the corresponding corrections to the self-energy of the $a_s$ transverse gauge field are proportional to $4\,e_s^2(\tilde{e}_c^2+\tilde{e}_s^2)$. It is instructive to contrast this with the hypothetical case in which both the fermion species

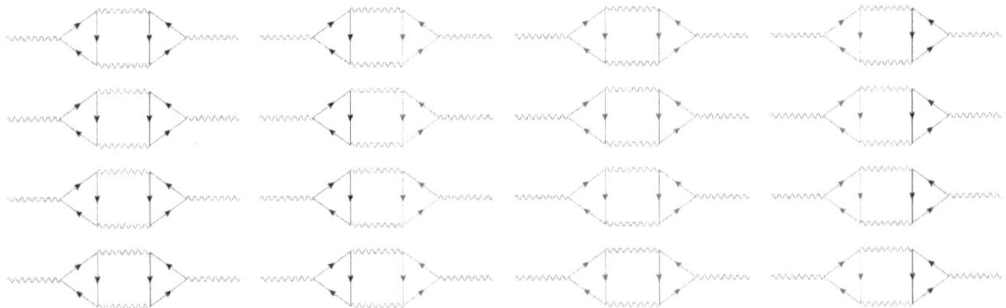

**Fig. 4.3** Aslamazov-Larkin-type diagrams contributing in the particle-particle channel to the three-loop bosonic self-energy of the $a_c$ transverse gauge field. Red and green wavy lines denote the $a_c$ and $a_s$ propagators, respectively, while black and cyan solid arrowed lines represent the $\psi_{1,\pm,j}$ and $\psi_{2,\pm,j}$ fermionic propagators, respectively. The combined contribution from all sixteen diagrams is proportional to $4\,e_c^2\,(\tilde{e}_c^2+\tilde{e}_s^2)$

carry identical charges under the two gauge fields: in that scenario, the contributions would instead be proportional to $4e_c^2(\tilde{e}_c+\tilde{e}_s)^2$ and $4e_s^2(\tilde{e}_c+\tilde{e}_s)^2$ for the $a_c$ and $a_s$ fields respectively, which would leave the fixed line intact. The fact that the actual contributions take a different form signals that the fixed line may be destabilized by three-loop corrections when $m=1$. For $m>1$, by contrast, the UV/IR mixing renders all higher-loop corrections $k_F$-suppressed, so they have no bearing on the fixed line. The fixed line feature is therefore generically robust against higher-loop corrections in this regime.

## 4.4 Conclusion

In this chapter, we have applied the dimensional regularization framework, developed for NFLs arising at the Ising-nematic quantum critical point (cf. Chapter 1), to the case of NFLs generated by transverse gauge-field couplings with finite-density fermions. This has allowed us to access the interacting fixed points perturbatively through an expansion in $\epsilon$, defined as the difference between the upper critical dimension, $d_c = m+3/(m+1)$, and the physical dimension $d_p = m+1$ for a FS of dimension $m$. The scaling behavior has been extracted for both the single $U(1)$ and the $U(1)\times U(1)$ gauge field cases.

A key distinction between the Ising-nematic and the gauge-field cases lies in the matrix structure of the couplings. This difference traces back to the fact that fermions at antipodal points on the FS couple to the Ising-nematic order parameter with the *same* sign, whereas they couple to a transverse gauge field with *opposite* signs. As a result, although the critical dimension and critical exponents turn out to be identical in the two cases, the differences surface in the renormalization of physical quantities such as the $2k_F$ scattering amplitudes—associated with backscattering processes carrying momentum $2k_F$—which can also be identified with a CDW instability. In particular, CDW ordering is *enhanced* near the NFL fixed point in the presence of transverse gauge field(s) for $m = 1$ [16], in contrast to the Ising-nematic scenario [10].

The $U(1) \times U(1)$ case is of especial interest in light of recent work showing that this framework provides a natural description of the deconfined Mott transition and the deconfined metal-metal transition [14]. Working in $(2+1)$ spacetime dimensions at one-loop order, Zou and Chowdhury found in Ref. [14] that these systems exhibit a continuous line of stable fixed points rather than a single isolated one. Their approach relied on modifying the bosonic dispersion to render it nonanalytic in momentum, followed by a double expansion in two small parameters [6, 7]. Our dimensional regularization scheme sidesteps this complication entirely, and carries the additional advantage of being applicable to FSs of generic dimension $m$ as well as being systematically improvable to higher loop orders, yielding corrections order by order in $\epsilon$. The discovery of a fixed line in Ref. [14] naturally raises the question of whether this feature persists at higher dimensions or beyond one loop. Our calculations demonstrate that extending to higher dimensions does not reduce the fixed line to a discrete set of fixed points, nor does it eliminate fixed points altogether.

As for higher-loop corrections, while we have not carried these out explicitly, arguments based on the known behavior at the Ising-nematic critical point [10, 11, 15] lead us to make informed predictions: While two-loop corrections will leave the fixed line intact, certain three-loop diagrams will destroy it. Should the fixed line be destroyed at three-loop order or beyond for $m = 1$, two scenarios could ensue: (1) the fixed line degenerates into a discrete set of fixed points, which may be stable or unstable; or (2) the beta-function develops no finite zeros, so that no fixed point exists at all. The physical consequences of each of these possibilities have been discussed in detail in Sect. V of Ref. [14].

## References


1. S. Chakravarty, R.E. Norton, O.F. Syljuåsen, Phys. Rev. Lett. **74**, 1423 (1995). https://doi.org/10.1103/PhysRevLett.74.1423
2. O.I. Motrunich, Phys. Rev. B **72**(4), 045105 (2005). https://doi.org/10.1103/PhysRevB.72.045105
3. S.S. Lee, P.A. Lee, Phys. Rev. Lett. **95**, 036403 (2005). https://doi.org/10.1103/PhysRevLett.95.036403. http://link.aps.org/doi/10.1103/PhysRevLett.95.036403
4. P.A. Lee, N. Nagaosa, X.G. Wen, Rev. Mod. Phys. **78**, 17 (2006). https://doi.org/10.1103/RevModPhys.78.17
5. O.I. Motrunich, M.P.A. Fisher, Phys. Rev. B **75**, 235116 (2007). https://doi.org/10.1103/PhysRevB.75.235116. http://link.aps.org/doi/10.1103/PhysRevB.75.235116
6. C. Nayak, F. Wilczek, Nucl. Phys. B **417**, 359 (1994). https://doi.org/10.1016/0550-3213(94)90477-4
7. D.F. Mross, J. McGreevy, H. Liu, T. Senthil, Phys. Rev. B **82**, 045121 (2010). https://doi.org/10.1103/PhysRevB.82.045121
8. S.B. Chung, I. Mandal, S. Raghu, S. Chakravarty, Phys. Rev. B **88**, 045127 (2013). https://doi.org/10.1103/PhysRevB.88.045127. http://link.aps.org/doi/10.1103/PhysRevB.88.045127

9. Z. Wang, I. Mandal, S.B. Chung, S. Chakravarty, Annals of Physics **351**, 727 (2014). https://doi.org/10.1016/j.aop.2014.09.021. http://www.sciencedirect.com/science/article/pii/S0003491614002814
10. D. Dalidovich, S.S. Lee, Phys. Rev. B **88**, 245106 (2013). https://doi.org/10.1103/PhysRevB.88.245106
11. I. Mandal, Eur. Phys. J. B **89**(12), 278 (2016). https://doi.org/10.1140/epjb/e2016-70509-4
12. I. Mandal, Phys. Rev. B **94**, 115138 (2016). https://doi.org/10.1103/PhysRevB.94.115138
13. M.A. Metlitski, S. Sachdev, Phys. Rev. B **82**, 075127 (2010). https://doi.org/10.1103/PhysRevB.82.075127
14. L. Zou, D. Chowdhury, Phys. Rev. Research **2**, 023344 (2020). https://doi.org/10.1103/PhysRevResearch.2.023344. https://link.aps.org/doi/10.1103/PhysRevResearch.2.023344
15. I. Mandal, S.S. Lee, Phys. Rev. B **92**, 035141 (2015). https://doi.org/10.1103/PhysRevB.92.035141
16. I. Mandal, Phys. Rev. Res. **2**, 043277 (2020). https://doi.org/10.1103/PhysRevResearch.2.043277. https://link.aps.org/doi/10.1103/PhysRevResearch.2.043277
17. S.S. Lee, Phys. Rev. B **78**(8), 085129 (2008). https://doi.org/10.1103/PhysRevB.78.085129
18. A. Eberlein, I. Mandal, S. Sachdev, Phys. Rev. B **94**, 045133 (2016). https://doi.org/10.1103/PhysRevB.94.045133. http://link.aps.org/doi/10.1103/PhysRevB.94.045133
19. S. Sachdev, *Quantum Phase Transitions*, 2nd edn. (Cambridge University Press, 2011). https://doi.org/10.1017/CBO9780511973765
20. B. Keimer, S.A. Kivelson, M.R. Norman, S. Uchida, J. Zaanen, Nature **518**(7538), 179 (2015). https://doi.org/10.1038/nature14165
21. T. Senthil, S. Sachdev, M. Vojta, Phys. Rev. Lett. **90**, 216403 (2003). https://doi.org/10.1103/PhysRevLett.90.216403. https://link.aps.org/doi/10.1103/PhysRevLett.90.216403

# Chapter 5 Non-Fermi Liquid Behaviour at the Onset of Incommensurate Charge-Density-Wave Order

## 5.1 Introduction

At quantum critical points (QCPs) hosting non-Fermi liquids (NFLs), the critical order-parameter bosons fall into two broad categories. In the first, the bosonic field is centered at zero momentum, causing quasiparticles to lose coherence across the entire FS [1–5]. In the second, the bosonic field is centered at a finite wavevector $\mathcal{Q}$ that connects points on the FS—commonly referred to as *hot-spots*—so that NFL behaviour emerges locally in their vicinity [1, 6–12]. The Ising-nematic critical point is a prominent example of the first category, while the second encompasses ordering transitions to phases such as spin-density wave (SDW), charge-density wave (CDW) [6–11, 13], and Fulde-Ferrell-Larkin-Ovchinnikov (FFLO) states [12].

CDW (SDW) bosons with $\mathcal{Q} \neq 0$ drive instabilities toward charge (magnetic) order, in which the charge (spin) density spontaneously breaks translational symmetry and develops a density modulation at wavevector $\mathcal{Q}$. These instabilities are further classified along two axes: whether the wavevector is commensurate or incommensurate with the underlying lattice, and whether $\mathcal{Q}$ constitutes a nesting vector of the FS. Commensurate wavevectors can be expressed as linear combinations of the reciprocal lattice vectors $\{\mathcal{R}\}$ with rational coefficients, while incommensurate ones cannot. Although there are infinitely many rational coefficients in principle, the quantitative effects of commensurability diminish as the size of their denominators grows. When $\mathcal{Q}$ coincides with a nesting vector connecting two points on the FS with antiparallel Fermi velocities (equivalently, antiparallel tangent vectors), the spin and charge orderings are enhanced by a well-known singularity arising from the enlarged phase space available for low-energy particle-hole excitations. In an inversion-symmetric crystal with valence band dispersion $\xi(\mathbf{k})$, the nesting vectors $\mathcal{Q}$ are determined by the condition $\xi(\mathcal{Q}/2 + \mathbf{G}/2) = \xi_{k_F}$, where $\xi_{k_F}$ is the Fermi energy. The nesting-vector condition, combined with inversion symmetry, implies $|\mathcal{Q}| = 2k_F$, where $k_F$ is the magnitude of the local Fermi momentum. This follows



I. Mandal, *Controlled Description Of Non-Fermi Liquids Using Field-Theoretic Methods*, SpringerBriefs in Physics, https://doi.org/10.1007/978-3-032-21618-2_5

from the fact that the two hot-spots are related by inversion symmetry and therefore share the same value of $k_F$.

The $2k_F$ wavevector instabilities are ubiquitous in two-dimensional systems displaying high-temperature superconductivity. Notable examples include the SDW instability at a $2k_F$ wavevector in the ground state of the 2d Hubbard model at half-filling [14, 15], and d-wave bond charge order, triggered by antiferromagnetic fluctuations, in models for cuprate superconductors [13, 16]. It is worth emphasizing that the nesting vector here causes only a *partial* nesting of the FS, which is conceptually distinct from perfect nesting, wherein entire slices—rather than discrete points—of the FS are connected by the same wavevector. Experimental evidence for incommensurate CDW ordering has been reported in a range of materials, including $NbSe_2$ and $TaS_2$ [17, 18], $VSe_2$ [19], $SmTe_3$ [20], and $TbTe_3$ [21]. In some of these compounds, the CDW transition temperature can be tuned toward zero by applying high pressure, revealing a putative QCP at the onset of CDW order [22]. These observations underscore the theoretical importance of understanding QCPs associated with incommensurate $2k_F$ wavevector instabilities.

In this chapter, we consider a pair of antipodal points on a one-dimensional FS of a two-dimensional (2d) metal, with parallel tangent vectors, interacting with an order-parameter boson whose condensation gives rise to an incommensurate CDW ordered phase [23, 24]. Right at the QCP, the CDW boson becomes massless, giving rise to strong quantum fluctuations that drive the system across a phase transition to an ordered state in which the electron density spontaneously breaks translational symmetry and develops a density modulation at a wavevector $\boldsymbol{Q}$ incommensurate with the reciprocal lattice. For definiteness, and without loss of generality, we take $\boldsymbol{Q} = 2k_F\,\hat{\boldsymbol{x}}$.

## 5.2 Model

Our starting point is the low-energy QFT action for finite-density fermions in two spatial dimensions, coupled to an incommensurate CDW order parameter centered at momentum $\boldsymbol{Q} = 2k_F\,\hat{\boldsymbol{x}}$. As shown in Fig. 5.1, the nesting vector $\boldsymbol{Q}$ connects a pair of hot-spots on the FS lying along the $x$-axis. The effective action governing the low-energy degrees of freedom near the hot-spots and the CDW order-parameter mode in $(2+1)$ dimensions reads [23, 24]

$$
\begin{aligned}
S = \sum_{s=\pm} \int_k \psi_s^\dagger(k)\left(-i\,k_0 + s\,k_1 + k_2^2\right)\psi_s(k) + \int_k \phi_+(k)\left(k_0^2 + k_1^2 + k_2^2\right)\phi_-(-k) \\
+ e \int_{k,q} \Big[\phi_+(q)\,\psi_+^\dagger(k+q)\,\psi_-(k) + \phi_-(-q)\,\psi_-^\dagger(k-q)\,\psi_+(k)\Big], \qquad (5.1)
\end{aligned}
$$

Here, $k = (k_0, \mathbf{k})$ denotes the three-vector comprising the Matsubara frequency $k_0$ and the spatial momentum $\mathbf{k} = (k_1, k_2) \equiv (k_x, k_y)$, with $\int_k \equiv \int dk_0\, d^d\mathbf{k}/(2\pi)^{d+1}$

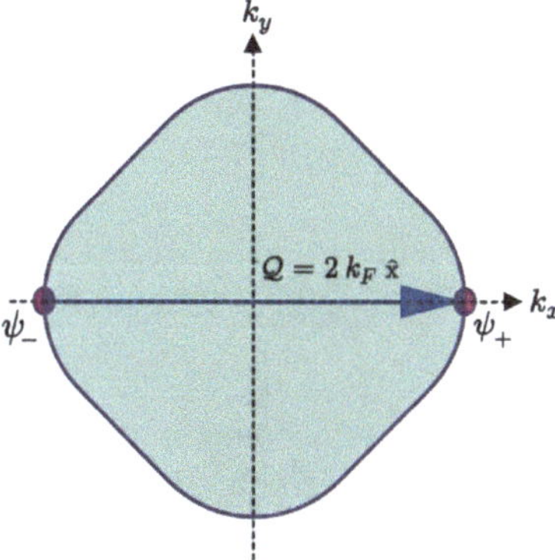


**Fig. 5.1** Schematics of the one-dimensional Fermi surface, showing the two hot-spots connected by the wavevector $\mathcal{Q} = 2k_F\,\hat{x}$ (blue arrow), which is incommensurate with the reciprocal lattice. The fermionic fields near the right and left hot-spots are labeled $\psi_+$ and $\psi_-$, respectively, and both couple to the CDW order-parameter bosonic fields with momenta centered at $\mathcal{Q}$

and $d = 2$ spatial dimensions. The fermionic degrees of freedom near the right and left hot-spots are represented by $\psi_+(k)$ and $\psi_-(k)$, respectively, while $\phi_+(k)$ and $\phi_-(k)$ denote the bosonic fluctuations carrying frequency $k_0$ and momenta $\mathcal{Q} + \mathbf{k}$ and $-\mathcal{Q} + \mathbf{k}$. The bosons are massless, reflecting the fact that we are working directly at the QCP. The fermionic momenta have been rescaled so that the Fermi velocity has unit magnitude and the FS curvature at the hot-spots equals 2. The bare bosonic velocity, although generically distinct from its fermionic counterpart, can be absorbed into a field redefinition and is accordingly set to unity. We further note that, at the critical point, the low-energy bosonic dynamics is dominated by particle-hole excitations of the FS, rendering the precise value of the bosonic velocity immaterial in the infrared (IR).

Since the FS is locally parabolic, we assign the scaling dimensions of 1 and $1/2$ to $k_1$ and $k_2$, respectively. To extract the critical scalings in a controlled way, we extend the co-dimension of the FS [2, 8, 25] to identify the upper critical dimension, $d = d_c$, at which the fermionic self-energy develops a logarithmic singularity. Maintaining analyticity in momentum space—or equivalently, locality in real space—for generic co-dimensions requires introducing the two-component spinors [2–4, 12, 26, 27],

$$\Psi(k) = \Big(\psi_+(k) \quad \psi_-^\dagger(-k)\Big)^T \quad \text{and} \quad \bar{\Psi} \equiv \Psi^\dagger\,\gamma_0\,. \tag{5.2}$$

Un terms of these spinor, the action describing the one-dimensional FS embedded in a $d$-dimensional momentum space, takes the form of

$$S = \int_k \bar{\Psi}(k)\, i\, (\mathbf{\Gamma} \cdot \mathbf{K} + \gamma_{d-1}\, \delta_k)\, \Psi(k) + \int_k \left(k_d^2 + \tilde{a}\, e_k\right) \phi_+(k)\, \phi_-(-k)$$
$$- \frac{i\, e\, \mu^{x/2}}{2} \int_{k,q} \left[\phi_+(q)\, \bar{\Psi}(k+q)\, \gamma_0\, \bar{\Psi}^T(-k) - \phi_-(-q)\, \Psi^T(q-k)\, \gamma_0\, \Psi(k)\right],$$
$$x = \frac{5}{2} - d\,, \quad \delta_k = k_{d-1} + k_d^2\,, \quad e_k = k_{d-1} + \frac{k_d^2}{2}\,. \tag{5.3}$$

The $(d-1)$-component vector, $\mathbf{K} \equiv (k_0, k_1, \ldots, k_{d-2})$, collects the frequency together with the $(d-2)$ momentum components arising from the added co-dimensions. The original momentum components along the $x$- and $y$-directions have been relabelled $k_{d-1}$ and $k_d$, respectively, so that within the $d$-dimensional momentum space, $\{k_1, \ldots, k_{d-1}\}$ spans the $(d-1)$ directions perpendicular to the FS while $k_d$ runs parallel to it. The matrix vector $\mathbf{\Gamma} \equiv (\gamma_0, \gamma_1, \ldots, \gamma_{d-2})$ carries $(d-1)$ components representing the gamma matrices associated with $k_0$ and the extra co-dimensions. Since the goal is ultimately to continue to $d = 2$, it suffices throughout to work with the $2 \times 2$ gamma matrices $\gamma_0 = \sigma_y$ and $\gamma_{d-1} = \sigma_x$.

In the purely bosonic sector, only the $k_d^2$ term in the kinetic energy survives, since $(|\mathbf{K}|^2 + k_{d-1}^2)$ is irrelevant under the patch-theory scaling [1–4, 12, 26], wherein each component of $\{\mathbf{K},\, k_{d-1}\}$ carries dimension one and $k_d$ carries dimension $1/2$. A dependence on $e_k$ in the bosonic propagator is generated dynamically through the susceptibility, driven by strong particle-hole fluctuations. We have therefore already included the term $\tilde{a}\, e_k$ in the action, whose form is fixed by the divergent contribution to the one-loop susceptibility derived below (see Eq. (5.12)). This term carries the same mass dimension as $k_d^2$, while $\tilde{a}$ itself has vanishing engineering dimension. Its omission would produce infrared divergences in loop integrals involving the bosonic propagator—spurious artifacts of truncating the effective action to the $k_d^2$ term alone. The engineering dimension of the fermion-boson coupling $e$ is $x/2$, which is why we have introduced an explicit factor of $\mu^{x/2}$: it renders $e$ dimensionless, as is standard practice in QFT calculations.

A noteworthy contrast with the Ising-nematic case (discussed in Chap. 2) is worth drawing here. In that setting, an emergent sliding symmetry [1–4] forces the terms proportional to $\bar{\Psi}(k)\, k_{d-1}\, \Psi(k)$ and $\bar{\Psi}(k)\, k_d^2\, \Psi(k)$ to renormalize in lockstep, so that the fermionic propagator depends on $k_{d-1}$ and $k_d^2$ only through the combination $\delta_k$, even after loop corrections are included. No such symmetry is operative in the present problem, and there is no a priori reason why sole dependence on $\delta_k$ should be preserved under renormalization. In principle, the corrected terms could conspire to flatten the FS at the hot-spots, as reported in the RPA calculations of Ref. [24]. As far as our explicit one-loop computations will show, however, no such flattening occurs at one-loop order of our QFT analysis. The flattening might show up when one computes higher-loop Feynman diagrams.

## 5.3 One-Loop Self-Energies and Implementation of Dimensional Regularization

The value of $x$ reveals that the coupling constant $e$ becomes marginal precisely at the upper critical dimension $d_c = 5/2$, remaining relevant for $d < 5/2$ and irrelevant for $d > 5/2$. We therefore seek to access the interacting phase through a perturbative expansion about $d = 5/2 - \epsilon$, where $\epsilon$ serves as the small control parameter, and the physical two-dimensional theory is ultimately recovered by setting $\epsilon = 1/2$. As a preparatory step toward deriving the RG flow equations, we compute the one-loop self-energies of the bosonic and fermionic sectors, which supply the key ingredients needed to determine the beta functions of the coupling constants $e$ and $\tilde{a}$. The bare fermionic and bosonic propagators, that follow from the action in Eq. (5.3), are

$$
\begin{aligned}
G_{(0)}(k) &\equiv \left\langle \Psi(k)\,\bar{\Psi}(k)\right\rangle_0 = \frac{1}{i}\,\frac{\mathbf{\Gamma}\cdot\mathbf{K} + \gamma_{d-1}\,\delta_k}{|\mathbf{K}|^2 + \delta_k^2}\,,\\
D^{+}_{(0)}(k) &\equiv \langle \phi_+(k)\,\phi_-(-k)\rangle_0 = \frac{1}{k_d^2 + \tilde{a}\,e_k}\,, \text{ and}\\
D^{-}_{(0)}(k) &\equiv \langle \phi_-(k)\,\phi_+(-k)\rangle_0 = D^{+}_{(0)}(k)\,.
\end{aligned}
\tag{5.4}
$$

### 5.3.1 *One-Loop Bosonic Self-Energy*

The one-loop bosonic self-energy (cf. Fig. 5.2a) is defined by

$$
\Pi(k) = -\frac{e^2\,\mu^x}{4} \times 2 \int_q \mathrm{Tr}\left[\gamma_0\, G_{(0)}(q)\,\gamma_0\, G^{T}_{(0)}(k-q)\right]. \tag{5.5}
$$

Applying the commutation relations between the gamma matrices alongside the identities $\gamma_{d-1}^T = -\gamma_0\,\gamma_{d-1}\,\gamma_0$ and $\Gamma^T = -\gamma_0\,\Gamma\,\gamma_0$, this simplifies to

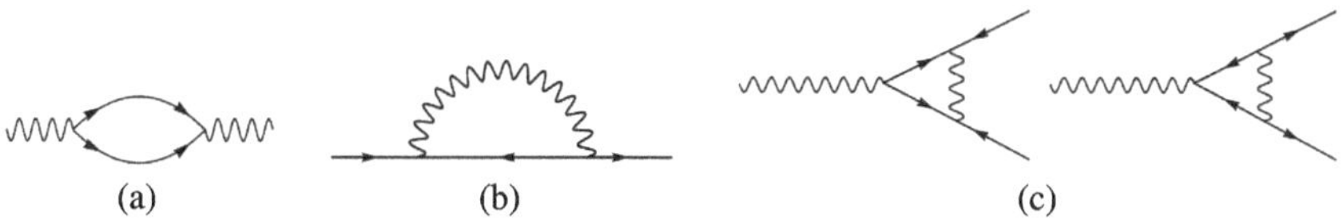


**Fig. 5.2** One-loop Feynman diagrams for the **a** bosonic self-energy, **b** fermionic self-energy, and **c** fermion-boson vertices. All the fermion propagators are represented by arrowed solid lines corresponding to the bare Green's function, $G_{(0)}$, while the dressed bosonic propagator, $D_{(1)}$, is depicted by wiggly lines

$$\Pi(k) = e^2\,\mu^x \int_q \frac{\mathbf{Q}\cdot(\mathbf{Q}-\mathbf{K}) - \delta_q\,\delta_{k-q}}{\left(\mathbf{Q}^2+\delta_q^2\right)\left[(\mathbf{Q}-\mathbf{K})^2+\delta_{k-q}^2\right]}\,. \tag{5.6}$$

Using $\delta_{k-q} = k_{d-1} + q_{d-1} + (k_d - q_d)^2$, we shift $q_{d-1} \to q_{d-1} - q_d^2$ and apply Feynman parametrization, arriving at

$$\begin{aligned}
\Pi(k) &= e^2\,\mu^x \int_q \int_0^1 dt\, \frac{|\mathbf{Q}|^2 - t(1-t)|\mathbf{K}|^2 - \tilde{e}_{kq}\,q_{d-1} + q_{d-1}^2}{\left[|\mathbf{Q}|^2 + t(1-t)|\mathbf{K}|^2 + t\,\tilde{e}_{kq}^2 + q_{d-1}^2 - 2t\,\tilde{e}_{kq}\,q_{d-1}\right]^2} \\
&= e^2\,\mu^x \int d^{d-1}|\mathbf{Q}|\,dq_d \int_0^1 dt\, \frac{|\mathbf{Q}|^d}{2^d\,\pi^{\frac{d+1}{2}}\,\Gamma\!\left(\frac{d-1}{2}\right)\left[|\mathbf{Q}|^2 + t(1-t)\left(\tilde{e}_{kq}^2 + |\mathbf{K}|^2\right)\right]^{3/2}} \\
&= e^2\,\mu^x \int dq_d\, \frac{2^{1-2d}\csc\!\left(\frac{d\pi}{2}\right)\left(\tilde{e}_{kq}^2 + |\mathbf{K}|^2\right)^{\frac{d-2}{2}}}{\pi^{\frac{d-1}{2}}\,\Gamma\!\left(\frac{d-1}{2}\right)}\,.
\end{aligned} \tag{5.7}$$

The substitution $u = \sqrt{2}\,q_d - k_d/\sqrt{2}$, with Jacobian $1/\sqrt{2}$, then gives

$$\Pi(k) = e^2\,\mu^x \int_0^\infty du\, \frac{2^{\frac{3}{2}-2d}\csc\!\left(\frac{d\pi}{2}\right)\left[\left(u^2+e_k\right)^2 + |\mathbf{K}|^2\right]^{\frac{d-2}{2}}}{\pi^{\frac{d-1}{2}}\,\Gamma\!\left(\frac{d-1}{2}\right)}\,. \tag{5.8}$$

Since the bare susceptibility diverges at zero temperature at the nesting vector $\mathcal{Q}$, a well-defined self-energy is obtained by subtracting off the singular zero-momentum contribution:

$$\tilde{\Pi}(k) = \Pi(k) - \Pi(0) = \frac{2^{\frac{3}{2}-2d}\csc\!\left(\frac{d\pi}{2}\right)e^2\,\mu^x}{\pi^{\frac{d-1}{2}}\,\Gamma\!\left(\frac{d-1}{2}\right)}\, I_\Pi(k,d)\,, \tag{5.9}$$

Upon the further substitution, $z = u^2 + e_k$, we get

$$\begin{aligned}
I_\Pi(k,d) &\equiv \int_{e_k}^\infty dz\, \frac{(z-e_k)^{2-d} - \left[z^2+|\mathbf{K}|^2\right]^{\frac{2-d}{2}}}{2\sqrt{z-e_k}\,\left[z^2+|\mathbf{K}|^2\right]^{\frac{2-d}{2}}\,(z-e_k)^{2-d}} \\
&= \begin{cases}
\frac{\Gamma(d-1)\,(-e_k)^{d-\frac{3}{2}}}{2}\left[\frac{\Gamma\left(\frac{3-d}{2}\right)\,{}_2F_1\left(\frac{3-2d}{4},\frac{5-2d}{4};\frac{3-d}{2};-\frac{|\mathbf{K}|^2}{e_k{}^2}\right)}{\sqrt{\pi}} + \frac{{}_2F_1\left(\frac{1}{2},1;d;1\right)}{\Gamma(d)}\right] \\
\quad + \frac{|\mathbf{K}|^d\,{}_3F_2\left(\frac{3}{4},1,\frac{5}{4};\frac{3}{2},\frac{d}{2}+1;-|\mathbf{K}|^2/e_k{}^2\right)}{4d\,(-e_k)^{3/2}} + \frac{\pi^{3/2}\,|\mathbf{K}|^{d-1}\sec\left(\frac{d\pi}{2}\right)\,{}_2F_1\left(\frac{1}{4},\frac{3}{4};\frac{d+1}{2};-|\mathbf{K}|^2/e_k^2\right)}{4\sqrt{-e_k}\,\Gamma\left(\frac{2-d}{2}\right)\Gamma\left(\frac{d+1}{2}\right)} \\
\quad + |\mathbf{K}|^{d-2}\sqrt{-e_k}\;{}_3F_2\left(\frac{1}{2},1,1-\frac{d}{2};\frac{3}{4},\frac{5}{4};-\frac{e_k^2}{|\mathbf{K}|^2}\right) + \frac{(-e_k)^{d-\frac{3}{2}}}{3-2d} & \text{for } e_k < 0 \\[2em]
\frac{\sqrt{\pi}\,e_k^{d-\frac{3}{2}}\,\Gamma\left(\frac{3-d}{2}\right)\,{}_2F_1\left(\frac{3-2d}{4},\frac{5-2d}{4};\frac{3-d}{2};-|\mathbf{K}|^2/e_k{}^2\right)}{2\,\Gamma(2-d)} & \text{for } e_k > 0\,.
\end{cases}
\end{aligned} \tag{5.10}$$

The zeros of $1/\Gamma\left(\frac{2-d}{2}\right)$ and $1/\Gamma(2-d)$ at $d=2$ are cancelled by the factor $\csc\left(\frac{d\pi}{2}\right)$ in Eq. (5.9), making overall term non-singular for $d=2$.

In the limit $|\mathbf{K}|^2/e_k^2 \ll 1$, the leading-order behaviour is

$$I_\Pi(k,d) = \begin{cases} \dfrac{\Gamma(d-1)(-e_k)^{d-\frac{3}{2}}}{2}\left[\dfrac{{}_2\tilde{F}_1\left(\frac{1}{2},1;d;1\right)}{\Gamma(d)} + \dfrac{\Gamma\left(\frac{3}{2}-d\right)}{\sqrt{\pi}} + \dfrac{\sqrt{\pi}}{\Gamma\left(d-\frac{1}{2}\right)}\right] \\ + \dfrac{(-e_k)^{d-\frac{3}{2}}}{3-2d} + \dfrac{\pi^{3/2}|\mathbf{K}|^{d-1}\sec\left(\frac{d\pi}{2}\right)}{2\sqrt{-e_k}\,\Gamma\left(\frac{2-d}{2}\right)\Gamma\left(\frac{d+1}{2}\right)} + \mathcal{O}\left(\dfrac{|\mathbf{K}|^d}{(-e_k)^{3/2}}\right) \text{ for } e_k < 0 \\ \\ \dfrac{\sqrt{\pi}\,\Gamma\left(\frac{3}{2}-d\right)e_k^{d-\frac{3}{2}}}{2\,\Gamma(2-d)} + \dfrac{\sqrt{\pi}\,\Gamma\left(\frac{7}{2}-d\right)|\mathbf{K}|^2}{4\,(d-3)\,\Gamma(2-d)\,e_k^{\frac{7-2d}{2}}} \\ + \mathcal{O}\left(\dfrac{|\mathbf{K}|^4}{e_k^{11/2-d}}\right) \text{ for } e_k > 0\,. \end{cases} \tag{5.11}$$

The poles of $\Gamma(3/2-d)$ and $1/(3-2d)$ at $d=3/2$ signal, respectively, a logarithmic divergence at $d=3/2$ and a linear divergence at $d=5/2$ in the Wilsonian language of a momentum cutoff $\Lambda \sim \mu$. Within dimensional regularization, UV divergences of all orders appear as poles of $\Gamma$-functions, and reinstating an explicit Wilsonian cutoff $\Lambda$ allows one to read off their degree. While these terms are important for the analysis of UV-stable fixed points, they must be set aside in the present context, as we are focused on the infrared RG flows in $d=5/2-\epsilon$ dimensions, where they correspond to IR-irrelevant operators. The situation is closely analogous to that of a $\phi^6$ interaction added to a $\phi^4$ scalar field theory in $(3+1)$ dimensions: the $\phi^4$ vertex is renormalizable with upper critical dimension 4, whereas the $\phi^6$ vertex has upper critical dimension 3, and its presence destroys renormalizability in four spacetime dimensions.

Expanding in $\epsilon$ about $d=5/2-\epsilon$ and dropping the terms with poles at $d=3/2$, we find

$$\left[\mu^x\,\frac{2^{\frac{3}{2}-2d}\csc\left(\frac{d\pi}{2}\right)}{\pi^{\frac{d-1}{2}}\,\Gamma\left(\frac{d-1}{2}\right)} \times I_\Pi(k,d)\right]\Bigg|_{d=\frac{5}{2}-\epsilon}$$

$$= \begin{cases} -\dfrac{\pi^{3/4}\,\dfrac{|\mathbf{K}|^{3/2}}{\sqrt{|e_k|}}\left(\dfrac{\mu}{|\mathbf{K}|}\right)^{\epsilon}}{32\sqrt{2}\,\Gamma^2(3/4)\,\Gamma(7/4)} + \mathcal{O}(\epsilon) \quad \text{for } e_k < 0 \\ -\dfrac{\dfrac{|\mathbf{K}|^2}{e_k}\left(\dfrac{\mu}{e_k}\right)^{\epsilon}}{32\pi^{3/4}\,\Gamma(3/4)} + \mathcal{O}(\epsilon) \qquad \text{for } e_k > 0\,. \end{cases}$$

The leading-order self-energy correction in Eq. (5.9) therefore takes the form,

$$\tilde{\Pi}(k) = -\frac{e^2\,\mu^x\,b\,|\mathbf{K}|^{d-1}}{\sqrt{|e_k|}}\,\Theta(-e_k)\,, \quad \text{where} \quad b = \frac{\pi^{3/4}}{32\sqrt{2}\,\Gamma^2(3/4)\,\Gamma(7/4)}\,, \tag{5.12}$$

in the limit $|\mathbf{K}|^2/e_k^2 \ll 1$.

Since the bare bosonic propagators $D^{\pm}_{(0)}(k)$ are independent of $\mathbf{K}$, loop integrals involving them are ill-defined without resumming a class of diagrams that endows the propagator with a nontrivial dispersion in these directions. We therefore dress the propagator by incorporating the finite one-loop correction $\tilde{\Pi}(k) \propto |\mathbf{K}|^{d-1}/\sqrt{|e_k|}$ in all subsequent loop calculations. For both $\phi_+(k)$ and $\phi_-(k)$, this amounts to working with

$$D_{(1)}(k) = \frac{1}{\left[D^{+}_{(0)}(k)\right]^{-1} - \tilde{\Pi}(k)} = \frac{1}{k_d^2 + \dfrac{b\,e^2\,\mu^x\,|\mathbf{K}|^{d-1}\,\Theta(-e_k)}{\sqrt{|e_k|}}}\,. \tag{5.13}$$

This reorganization of the perturbative expansion ensures that the $\mathbf{K}$-dependent finite part of the one-loop bosonic self-energy is already accounted for at the zeroth order. The term $\tilde{\Pi}(k)$ is the well-known *Landau-damping* contribution, which underlies the characteristic $\text{sgn}(k_0)|k_0|^{2/3}$ frequency dependence of the fermionic self-energy—a robust signature of NFL behaviour that has been identified at the QCPs in a broad range of strongly-correlated systems [1–4, 12, 13, 26].

### 5.3.2 *One-Loop Fermion Self-Energy*

Turning to the fermionic sector, the one-loop self-energy (cf. Fig. 5.2b) is expressed as the integral

$$\Sigma(k) = e^2\,\mu^x \int_q \gamma_0\,G^T_{(0)}(q-k)\,\gamma_0\,D_{(1)}(q) = i\,\Sigma_1(k)\,\Gamma\cdot\mathbf{K} + i\,\Sigma_2(k)\,\gamma_{d-1}\,, \tag{5.14}$$

where

$$\Sigma_1(k) = -\frac{e^2\,\mu^x}{|\mathbf{K}|^2}\int_q \frac{\mathbf{K}\cdot(\mathbf{Q}-\mathbf{K})}{(\mathbf{Q}-\mathbf{K})^2 + \delta^2_{q-k}}\,D_{(1)}(q) \tag{5.15}$$

and

$$\Sigma_2(k) = e^2\,\mu^x \int_q \frac{\delta_{q-k}}{(\mathbf{Q}-\mathbf{K})^2 + \delta_{q-k}^2}\, D_{(1)}(q)\,. \tag{5.16}$$

The evaluation of these two contributions is detailed in the following two subsections, which the reader may wish to skip if they prefer to bypass the more involved intermediate steps. For convenience, we collect the final results here. Setting $d = d_c - \epsilon$, the singular part evaluates to

$$\Sigma(k) = -\frac{e^{4/3}\,\mathcal{U}_1}{(2-\tilde{a})^{2/3}\,\epsilon}\, i\,(\Gamma \cdot \mathbf{K}) + \mathcal{O}\epsilon^0, \quad \mathcal{U}_1 = \frac{\sqrt{2}\,\Gamma\left(\frac{5}{4}\right)}{3\,\sqrt{3}\,\pi^{7/4}\,b^{1/3}}\,, \tag{5.17}$$

where the logarithmic divergence is reflected by a pole at $\epsilon = 0$.

#### 5.3.2.1 Computation of Γ-Dependent Part

The leading-order dependence of $\Sigma_1(k)$ on $\mathbf{K}$ is captured by setting the external momentum components $k_d$ and $k_{d-1}$ to zero, reducing the problem to evaluating

$$\begin{aligned}
&\Sigma_1(\mathbf{K},0,0)\\
&= \frac{e^2\,\mu^x}{|\mathbf{K}|^2}\int_q \frac{\mathbf{K}\cdot(\mathbf{K}-\mathbf{Q})}{(\mathbf{Q}-\mathbf{K})^2+\delta_q^2} \times \frac{1}{q_d^2 + e^2\,\mu^x\, b\,|\mathbf{Q}|^{d-1}\,\Theta(-e_q)/\sqrt{|e_q|}}\,.
\end{aligned} \tag{5.18}$$

Switching to $q_d$ and $e_q$ as integration variables, and splitting the integration domain into the regions $e_q < 0$ and $e_q > 0$ via $\Sigma_1(\mathbf{K},0,0) = I_1 + I_2$, we obtain

$$\begin{aligned}
I_1 &= \frac{e^2\,\mu^x}{|\mathbf{K}|^2}\int_{e_q<0} \frac{d^{d-1}\mathbf{Q}\,dq_d\,de_q}{(2\,\pi)^{d+1}}\, \frac{-\mathbf{K}\cdot(\mathbf{Q}-\mathbf{K})}{(\mathbf{Q}-\mathbf{K})^2+\left(e_q+q_d^2/2\right)^2}\, \frac{1}{q_d^2+e^2\,\mu^x\,b\,|\mathbf{Q}|^{d-1}/\sqrt{|e_q|}}\\
&= \frac{e^2\,\mu^x}{|\mathbf{K}|^2}\int_0^\infty \frac{du}{\sqrt{u/2}}\int_0^\infty de_q \int_{-\infty}^\infty \frac{d^{d-1}\mathbf{Q}}{(2\,\pi)^{d+1}}\, \frac{\mathbf{K}^2-\mathbf{K}\cdot\mathbf{Q}}{(\mathbf{Q}-\mathbf{K})^2+\left(u-e_q\right)^2}\\
&\qquad\qquad \times \frac{1}{2\,u+e^2\,\mu^x\,b\,|\mathbf{Q}|^{d-1}/\sqrt{e_q}}
\end{aligned} \tag{5.19}$$

and

$$\begin{aligned}
I_2 &= \frac{e^2\,\mu^x}{|\mathbf{K}|^2}\int_{e_q>0} \frac{d^{d-1}\mathbf{Q}\,dq_d\,de_q}{(2\,\pi)^{d+1}}\, \frac{-\mathbf{K}\cdot(\mathbf{Q}-\mathbf{K})}{(\mathbf{Q}-\mathbf{K})^2+\left(e_q+q_d^2/2\right)^2}\, \frac{1}{q_d^2}\\
&= \frac{e^2\,\mu^x}{|\mathbf{K}|^2}\int_{e_q>0} \frac{d^{d-1}\mathbf{Q}\,dq_d\,de_q}{(2\,\pi)^{d+1}}\, \frac{-\mathbf{K}\cdot\mathbf{Q}}{\mathbf{Q}^2+\left(e_q+q_d^2/2\right)^2}\, \frac{1}{q_d^2} = 0\,.
\end{aligned} \tag{5.20}$$

The integral $I_1$ does not admit an exact closed form, and we proceed by identifying the regions that dominate the integrand. The first factor concentrates weight near $|\mathbf{Q}| \sim |\mathbf{K}|$ and $u \sim e_q$, while the second factor is dominated by $e_q \sim |\mathbf{Q}|^{2(d-1)/3} \sim$

$|\mathbf{K}|^{2(d-1)/3}$. Since $|\mathbf{K}|^{2(d-1)/3} \gg |\mathbf{K}|$ for small $|\mathbf{K}|$ when $2(d-1)/3 < 1$, we may replace $u$ by $e_q$ in both the $\sqrt{u}$ factor in the overall denominator and the $2u$ term in the denominator of the second factor. Extending the lower limit of the $u$-integration to $-\infty$ then yields

$$
\begin{aligned}
I_1 &\simeq \frac{e^2\,\mu^x}{|\mathbf{K}|^2}\int_{-\infty}^{\infty}\frac{d^{d-1}\mathbf{Q}\,du}{(2\pi)^{d+1}}\int_{e_q>0}\frac{de_q}{\sqrt{e_q/2}}\,\frac{\mathbf{K}\cdot(\mathbf{K}-\mathbf{Q})}{(\mathbf{Q}-\mathbf{K})^2+u^2}\,\frac{1}{2\,e_q+e^2\,\mu^x\,b\,|\mathbf{Q}|^{d-1}/\sqrt{e_q}} \\
&= \frac{e^2\,\mu^x}{|\mathbf{K}|^2}\int_{-\infty}^{\infty}\frac{d^{d-1}\mathbf{Q}\,du}{(2\pi)^{d+1}}\int_{e_q>0}de_q\,\frac{\mathbf{K}\cdot(\mathbf{K}-\mathbf{Q})}{(\mathbf{Q}-\mathbf{K})^2+u^2}\,\frac{\sqrt{2}}{2\,e_q^{3/2}+e^2\,\mu^x\,b\,|\mathbf{Q}|^{d-1}} \\
&= -\frac{e^{4/3}\,\Gamma\left(\frac{5-2d}{6}\right)\Gamma\left(\frac{d}{2}\right)\Gamma\left(\frac{d+2}{6}\right)}{2^{\frac{4d-1}{6}}\,\pi^{\frac{d+1}{2}}\times 3\sqrt{3}\times 2^{2/3}\,b^{1/3}\,\Gamma\left(\frac{5d-2}{6}\right)}\left(\frac{\mu}{|\mathbf{K}|}\right)^{\frac{2x}{3}}.
\end{aligned}
\tag{5.21}
$$

The integral diverges at $d = 5/2$, thereby confirming this as the value of the upper critical dimension, $d_c$. The fermion-boson coupling, $e$, is irrelevant for $d > d_c$ and relevant for $d < d_c$, and marginal precisely at $d = d_c$. This structure opens the door to a controlled perturbative treatment of the strongly interacting NFL state by working in $d = 5/2 - \epsilon$, where $\epsilon$ plays the role of a small expansion parameter. Within our dimensional regularization scheme, the divergence manifests as a $\sim \epsilon^{-1}$ pole, arising from the factor $\Gamma\left(\frac{5-2d}{6}\right)$ at $d = d_c$. We further note that this term reproduces the fermionic self-energy behavior $\sim \mathrm{sgn}(k_0)|k_0|^{2/3}$ at $d = 2$, in agreement with the uncontrolled RPA result [23, 24]. It is worth emphasizing that this correct $k_0$-dependence of $\Sigma$ could only be captured by incorporating the Landau-damping term into the dressed bosonic propagator $D_{(1)}$ from the outset.

#### 5.3.2.2 Computation of $\gamma_{d-1}$-Dependent Part

The leading dependence of $\Sigma_2(k)$ on $k_d$ and $k_{d-1}$ is captured by setting $\mathbf{K} = 0$, reducing the problem to evaluating

$$
\Sigma_2(\mathbf{0}, k_d, k_{d-1}) = e^2\,\mu^x e^2\,\mu^x\left[I_3(k_d, k_{d-1}) + I_4(k_d, k_{d-1})\right], \tag{5.22}
$$

where $\delta_{q-k} = q_{d-1} - k_{d-1} + k_d^2 + q_d^2 - 2\,k_q\,q_d$ and

$$
\begin{aligned}
I_3(k_d, k_{d-1}) &= \int_{q,\,e_q<0}\frac{\delta_{q-k}}{|\mathbf{Q}|^2+\delta_{q-k}^2}\times\frac{1}{q_d^2+e^2\,\mu^x\,b\,|\mathbf{Q}|^{d-1}/\sqrt{|e_q|}} \quad\text{and} \\
I_4(k_d, k_{d-1}) &= \int_{q,\,e_q>0}\frac{\delta_{q-k}}{|\mathbf{Q}|^2+\delta_{q-k}^2}\times\frac{1}{q_d^2}.
\end{aligned}
\tag{5.23}
$$

To streamline the calculation, we set $k_d = 0$, whereupon

$$
\begin{aligned}
I_3(0, k_{d-1}) &= \int_{q,\, e_q>0} \frac{q_d^2/2 - e_q - k_{d-1}}{|\mathbf{Q}|^2 + \left(q_d^2/2 - e_q - k_{d-1}\right)^2} \times \frac{1}{q_d^2 + e^2\, \mu^x\, b\, |\mathbf{Q}|^{d-1}/\sqrt{e_q}} \\
&= \int_0^\infty \frac{du}{\sqrt{u/2}} \int_0^\infty de_q \int_{-\infty}^\infty \frac{d^{d-1}\mathbf{Q}}{(2\,\pi)^{d+1}} \frac{u - e_q - k_{d-1}}{|\mathbf{Q}|^2 + \left(u - e_q - k_{d-1}\right)^2} \\
&\quad \times \frac{1}{2\,u + e^2\, \mu^x\, b\, |\mathbf{Q}|^{d-1}/\sqrt{e_q}} \quad \left(\text{where } 2\,u = q_d^2\right). \qquad (5.24)
\end{aligned}
$$

The first factor of the integrand concentrates the dominant contribution near $|\mathbf{Q}| \sim 0$ and $u \sim e_q + k_{d-1}$, while the second factor is dominated by $e_q \sim |\mathbf{Q}|^{2(d-1)/3}$. We may therefore replace $u$ by $e_q + k_{d-1}$ in both the $\sqrt{u}$ factor in the overall denominator and the $2u$ term in the denominator of the second factor, and extend the lower limit of the $u$-integration to $-\infty$. This yields

$$
\begin{aligned}
I_3(0, k_{d-1}) &\simeq \int_0^\infty \frac{du}{\sqrt{e_q + k_{d-1}}} \int_0^\infty de_q \int_{-\infty}^\infty \frac{d^{d-1}\mathbf{Q}}{(2\,\pi)^{d+1}} \frac{u - e_q - k_{d-1}}{|\mathbf{Q}|^2 + \left(u - e_q - k_{d-1}\right)^2} \\
&\quad \times \frac{\sqrt{2}}{2\,e_q + 2\,k_{d-1} + e^2\, \mu^x\, b\, |\mathbf{Q}|^{d-1}/\sqrt{e_q}} = 0\,. \qquad (5.25)
\end{aligned}
$$

To leading order, the integral $I_3$ therefore vanishes, leaving no $b$-dependent contribution. Proceeding to the next term, we have

$$
\begin{aligned}
I_4(0, k_{d-1}) &= \int_{q,\, e_q>0} \frac{q_d^2/2 + e_q - k_{d-1}}{|\mathbf{Q}|^2 + \left(q_d^2/2 + e_q - k_{d-1}\right)^2} \times \frac{1}{q_d^2} \\
&= \int_0^\infty \frac{du}{\sqrt{u/2}} \int_{-\infty}^0 de_q \int_{-\infty}^\infty \frac{d^{d-1}\mathbf{Q}}{(2\,\pi)^{d+1}} \frac{u + e_q - k_{d-1}}{|\mathbf{Q}|^2 + \left(u + e_q - k_{d-1}\right)^2} \frac{1}{2\,u}\,. \qquad (5.26)
\end{aligned}
$$

Applying the same reasoning as before, the first factor of the integrand concentrates the dominant contribution near $|\mathbf{Q}| \sim 0$ and $u \sim k_{d-1} - e_q$. Substituting $u \sim k_{d-1} - e_q$ into the $\sqrt{u}$ factor in the overall denominator and the $2u$ term in the denominator of the second factor, and extending the lower limit of the $u$-integration to $-\infty$, we find $I_4(0, k_{d-1}) = 0$. It follows that the FS flattening at the hot-spots, which appears in the RPA calculations of Ref. [24], does not emerge in our one-loop computation.

### 5.3.3 One-Loop Vertex Correction

The one-loop vertex corrections (cf. Fig. 5.2c) are finite and therefore play no role in the renormalization procedure or the resulting RG flows.

## 5.4 RG Flows Under the Minimal Subtraction Scheme

The counterterm action, constructed to absorb the singular contributions, takes the form

$$S_{CT} = \int_k \bar{\Psi}(k)\, i \left[ A_1 \,\mathbf{\Gamma} \cdot \mathbf{K} + \gamma_{d-1} \left( A_2\, e_k + A_3 \frac{k_d^2}{2} \right) \right] \Psi(k) + \int_k A_4\, k_d^2\, \phi_+(k)\, \phi_-(-k)$$
$$- \frac{i\, e\, \mu^{x/2}}{2} \int_{k,q} A_6 \left[ \phi_+(q)\, \bar{\Psi}(k+q)\, \gamma_0\, \bar{\Psi}^T(-k) - \phi_-(-q)\, \Psi^T(q-k)\, \gamma_0\, \Psi(k) \right], \tag{5.27}$$

where $A_\zeta = \sum_{n=1}^{\infty} Z_\zeta^{(n)}/\epsilon^n$ with $\zeta \in [1,5]$. The $(d-1)$-dimensional rotational invariance in the space perpendicular to the FS ensures that every term in $\mathbf{\Gamma} \cdot \mathbf{K}$ is renormalized in the same way.

Subtracting $S_{CT}$ from the *bare* action $S_{\text{bare}}$ yields the renormalized action, which constitutes the *physical* effective action of the theory, expressed entirely in terms of non-divergent quantum parameters. While the bare parameters may be divergent, the physical observables are identified with the renormalized coupling constants, whose evolution is governed by the RG equations. These describe how the couplings flow as functions of the floating energy scale $\mu\, e^{-l}$, or equivalently, as the logarithmic length scale $l$ increases. To set this up, we first introduce the bare action,

$$S_{\text{bare}} = \int_{k^B} \bar{\Psi}^B(k^B)\, i \left[ \mathbf{\Gamma} \cdot \mathbf{K}^B + \gamma_{d-1} \left\{ e_k^B + \frac{\left(k^B\right)^2}{2} \right\} \right] \Psi^B(k^B)$$
$$+ \int_{k^B} \left(k_d^B\right)^2 \phi_+^B(k^B)\, \phi_-^B(-k^B)$$
$$- \frac{i\, e^B}{2} \int_{k^B,\, q^B} \Big[ \phi_+^B(q^B)\, \bar{\Psi}^B(k^B+q^B)\, \gamma_0 \left(\bar{\Psi}^B(-k^B)\right)^T$$
$$- \phi_-^B(-q^B) \left(\Psi^B(q^B-k^B)\right)^T \gamma_0\, \Psi^B(k^B) \Big], \tag{5.28}$$

comprising of the *bare quantities*. The superscript "$B$" labels bare fields, couplings, frequencies, and momenta throughout. The bare quantities are related to their renormalized counterparts (those without the superscript "$B$") through the multiplicative $Z_\zeta$-factors, via

$$S_{\text{bare}} = S + S_{CT}\,, \quad Z_\zeta = 1 + A_\zeta\,, \quad \mathbf{K}^B = \frac{Z_1}{Z_3}\,\mathbf{K}\,, \quad e_k^B = \frac{Z_2}{Z_3}\,e_k\,, \quad k_d^B = k_d\,,$$
$$\Psi^B(k^B) = Z_\Psi^{1/2}\,\Psi(k)\,, \quad \phi_\pm^B(k^B) = Z_\phi^{1/2}\,\phi_\pm\,, \tag{5.29}$$

and

$$Z_\Psi = Z_1 \left(\frac{Z_1}{Z_3}\right)^{-d} \left(\frac{Z_2}{Z_3}\right)^{-1}, \quad Z_\phi = Z_4 \left(\frac{Z_1}{Z_3}\right)^{1-d} \left(\frac{Z_2}{Z_3}\right)^{-1},$$
$$e^B = Z_e\, e\, \mu^{\frac{\epsilon}{2}}\,, \quad Z_e = \frac{Z_5 \left(\frac{Z_1}{Z_3}\right)^{1-\frac{d}{2}} \left(\frac{Z_2}{Z_3}\right)^{-1/2}}{\sqrt{Z_1}\, Z_4}\,. \tag{5.30}$$

There exists a freedom to rescale both fields and momenta without affecting the action, and we exploit this by fixing $k_d^B = k_d$, which amounts to measuring the scaling dimensions of all other quantities relative to that of $k_d$. The resulting $S$ is the renormalized—or Wilsonian effective—action, expressed entirely in terms of renormalized quantities. In essence, we have written the fundamental action of the theory in two equivalent ways, which allows the divergent contributions collected in $S_{CT}$ to be systematically subtracted off.

### 5.4.1 RG Flow Equations from the One-Loop Results

At one-loop order, the divergent contributions are extracted from Eqs. (5.12) and (5.17), and yield

$$Z_1 = 1 - \frac{e^{4/3}\,\mathcal{U}_1}{2^{2/3}\,\epsilon}\,, \quad Z_2 = Z_3 = Z_4 = Z_5 = 1\,,$$
$$b = \frac{\pi^{3/4}}{32\sqrt{2}\,\Gamma^2\!\left(\frac{3}{4}\right)\Gamma\!\left(\frac{7}{4}\right)}\,, \quad \mathcal{U}_1 = \frac{\sqrt{2}\,\Gamma\!\left(\frac{5}{4}\right)}{3\sqrt{3}\,\pi^{7/4}\,b^{1/3}}\,. \tag{5.31}$$

At this order, we find that $Z_2 = Z_3$, with neither receiving any correction from the loop integrals.

Since $Z_2 = Z_3$, a single dynamical critical exponent suffices for the fermions,

$$z = 1 + \frac{\partial \ln\!\left(\frac{Z_1}{Z_2}\right)}{\partial \ln \mu} = 1 + \frac{\partial \ln\!\left(\frac{Z_1}{Z_3}\right)}{\partial \ln \mu}\,, \tag{5.32}$$

reflecting the fact that the $\delta_k$ term, taken as a whole, is not renormalized at one-loop order. The anomalous dimensions of the fermions and bosons are defined by

$$\eta_\psi = \frac{1}{2}\frac{\partial \ln Z_\psi}{\partial \ln \mu} \quad \text{and} \quad \eta_\phi = \frac{1}{2}\frac{\partial \ln Z_\phi}{\partial \ln \mu}\,, \tag{5.33}$$

respectively, and the beta function for $e$ is

$$\beta_e = \frac{de}{d\ln\mu}\,. \tag{5.34}$$

The mass scale $\mu$ was introduced purely as a regularization device to render the loop-integrals finite. Since it is not a parameter of the fundamental theory, physical observables must be independent of it, and the same must hold for the bare parameters. Imposing this requirement, together with the condition that the non-singular parts of the solutions admit the small-$\epsilon$ expansions

$$z = z^{(0)}\,, \quad \eta_\psi = \eta_\psi^{(0)} + \eta_\psi^{(1)}\,\epsilon\,, \quad \eta_\phi = \eta_\phi^{(0)} + \eta_\phi^{(1)}\,\epsilon\,, \quad \beta_e = \beta_e^{(0)} + \beta_e^{(1)}\,\epsilon\,, \tag{5.35}$$

in the limit $\epsilon \to 0$, one arrives at the differential equations,

$$\begin{aligned} &z = 1 + \beta_e^{(1)}\,\frac{\partial Z_1^{(1)}}{\partial e}\,, \quad \eta_\psi = \frac{1}{4}\left(5 - 5z + 2\,\frac{\partial Z_1^{(1)}}{\partial e}\,\beta_e^{(1)}\right) + \frac{(z-1)\,\epsilon}{2}\,, \\ &\eta_\phi = \frac{3-3z}{4} + \frac{(z-1)\,\epsilon}{2}\,, \quad \frac{4\,\beta_e^{(0)}}{e} = -e\,z\,\frac{\partial Z_1^{(1)}}{\partial e} + z - 1\,, \quad \beta_e^{(1)} = -\frac{e\,z}{2}\,. \end{aligned} \tag{5.36}$$

These equations are derived by (1) imposing $\frac{d}{d\ln\mu}$(bare quantity) $= 0$; (2) substituting the expressions from Eqs. (5.31) and (5.35); (3) expanding in powers of $\epsilon$; and (4) matching coefficients of regular powers of $\epsilon$ on both sides. Solving this system yields

$$-\frac{\beta_e}{e} = \frac{6\epsilon - 3\times 2^{1/3}\,\mathcal{U}_1\,\tilde{e}}{12 - 2^{7/3}\,\mathcal{U}_1\,\tilde{e}}\,, \quad z = \frac{3}{3 - 2^{1/3}\,\mathcal{U}_1\,\tilde{e}}\,, \quad \eta_\psi = \eta_\phi = \frac{(3-2\,\epsilon)\,\mathcal{U}_1\,\tilde{e}}{4\mathcal{U}_1\,\tilde{e} - 6\times 2^{2/3}}\,, \tag{5.37}$$

where $\tilde{e} = e^{4/3}$. Since we are interested in the behavior at infrared energy scales, we track the RG flows with respect to the logarithmic length scale $l$, through the derivative

$$\frac{de}{dl} \equiv -\beta_e \tag{5.38}$$

for the coupling constant $e$.

### 5.4.2 *Stability of the Fixed Points of the RG Flows*

The fixed points of the RG-flow differential equation are located where the beta-function, $\beta_e$, vanishes. They are readily found to be $e = 0$ and $\tilde{e} = 2^{2/3}\,\epsilon/\mathcal{U}_1$. To determine the stability of each fixed point, one examines whether the RG-flow lines in the IR, generated by $\{-\beta_e\}$, flow toward or away from it. This classifies the fixed point as stable or unstable, accordingly. For $\epsilon \in (0, 1/2]$, there is precisely one stable interacting fixed point for each value of $\epsilon$, located at $\tilde{e} = 2^{2/3}\,\epsilon/\mathcal{U}_1$, while $e = 0$ corresponds to an unstable Gaussian (non-interacting) fixed point. At the stable fixed point, the critical exponents evaluate to $z = 1 + 2\,\epsilon/3$ and $\eta_\phi = \eta_\psi = -\,\epsilon/2$.

## 5.5 Conclusion

This chapter has been devoted to a QFT analysis of the QCP that governs the continuous phase transition between a normal metallic state and a phase exhibiting incommensurate CDW order. A key ingredient throughout has been the use of a bosonic propagator augmented by the Landau-damping correction $\Pi_{\rm LD}$, which proves essential in generating the NFL fixed point. In particular, this dressed propagator is responsible for causing the frequency-dependence $\sim \text{sgn}(k_0)|k_0|^{2/3}$ appearing in the fermionic self-energy, that is a defining characteristic of an NFL behaviour [2–4, 12, 26–28].

## References

1. M.A. Metlitski, S. Sachdev, Phys. Rev. B **82**, 075127 (2010). https://doi.org/10.1103/PhysRevB.82.075127
2. D. Dalidovich, S.S. Lee, Phys. Rev. B **88**, 245106 (2013). https://doi.org/10.1103/PhysRevB.88.245106
3. I. Mandal, S.S. Lee, Phys. Rev. B **92**, 035141 (2015). https://doi.org/10.1103/PhysRevB.92.035141
4. I. Mandal, Eur. Phys. J. B **89**(12), 278 (2016). https://doi.org/10.1140/epjb/e2016-70509-4
5. I. Mandal, Phys. Rev. B **94**, 115138 (2016). https://doi.org/10.1103/PhysRevB.94.115138
6. A. Abanov, A. Chubukov, Phys. Rev. Lett. **93**(25), 255702 (2004). https://doi.org/10.1103/PhysRevLett.93.255702
7. A. Abanov, A.V. Chubukov, Phys. Rev. Lett. **84**, 5608 (2000). https://doi.org/10.1103/PhysRevLett.84.5608
8. S. Sur, S.S. Lee, Phys. Rev. B **91**, 125136 (2015). https://doi.org/10.1103/PhysRevB.91.125136
9. I. Mandal, Ann. Phys. **376**, 89 (2017). https://doi.org/10.1016/j.aop.2016.11.009. http://www.sciencedirect.com/science/article/pii/S0003491616302585
10. A. Schlief, P. Lunts, S.S. Lee, Phys. Rev. X **7**, 021010 (2017). https://doi.org/10.1103/PhysRevX.7.021010. https://link.aps.org/doi/10.1103/PhysRevX.7.021010
11. P. Lunts, A. Schlief, S.S. Lee, Phys. Rev. B **95**, 245109 (2017). https://doi.org/10.1103/PhysRevB.95.245109. https://link.aps.org/doi/10.1103/PhysRevB.95.245109

12. D. Pimenov, I. Mandal, F. Piazza, M. Punk, Phys. Rev. B **98**, 024510 (2018). https://doi.org/10.1103/PhysRevB.98.024510
13. M.A. Metlitski, S. Sachdev, Phys. Rev. B **82**, 075128 (2010). https://doi.org/10.1103/PhysRevB.82.075128
14. H.J. Schulz, Phys. Rev. Lett. **64**, 1445 (1990). https://doi.org/10.1103/PhysRevLett.64.1445. https://link.aps.org/doi/10.1103/PhysRevLett.64.1445
15. P.A. Igoshev, M.A. Timirgazin, A.A. Katanin, A.K. Arzhnikov, V.Y. Irkhin, Phys. Rev. B **81**, 094407 (2010). https://doi.org/10.1103/PhysRevB.81.094407. https://link.aps.org/doi/10.1103/PhysRevB.81.094407
16. S. Sachdev, R. La Placa, Phys. Rev. Lett. **111**, 027202 (2013). https://doi.org/10.1103/PhysRevLett.111.027202. https://link.aps.org/doi/10.1103/PhysRevLett.111.027202
17. F.D.S. J.A. Wilson, S. Mahajan, Adv. Phys. **24**(2), 117 (1975). https://doi.org/10.1080/00018737500101391
18. G. Scholz, O. Singh, R. Frindt, A. Curzon, Solid State Commun. **44**(10), 1455 (1982). https://doi.org/10.1016/0038-1098(82)90030-8. https://www.sciencedirect.com/science/article/pii/0038109882900308
19. P. Chen, W.W. Pai, Y.H. Chan, V. Madhavan, M.Y. Chou, S.K. Mo, A.V. Fedorov, T.C. Chiang, Phys. Rev. Lett. **121**, 196402 (2018). https://doi.org/10.1103/PhysRevLett.121.196402. https://link.aps.org/doi/10.1103/PhysRevLett.121.196402
20. G.H. Gweon, J.D. Denlinger, J.A. Clack, J.W. Allen, C.G. Olson, E. DiMasi, M.C. Aronson, B. Foran, S. Lee, Phys. Rev. Lett. **81**, 886 (1998). https://doi.org/10.1103/PhysRevLett.81.886. https://link.aps.org/doi/10.1103/PhysRevLett.81.886
21. A. Fang, N. Ru, I.R. Fisher, A. Kapitulnik, Phys. Rev. Lett. **99**, 046401 (2007). https://doi.org/10.1103/PhysRevLett.99.046401. https://link.aps.org/doi/10.1103/PhysRevLett.99.046401
22. Y. Feng, J. van Wezel, J. Wang, F. Flicker, D.M. Silevitch, P.B. Littlewood, T.F. Rosenbaum, Nat. Phys. **11**(10), 865 (2015). https://doi.org/10.1038/nphys3416
23. T. Holder, W. Metzner, Phys. Rev. B **90**, 161106 (2014). https://doi.org/10.1103/PhysRevB.90.161106. https://link.aps.org/doi/10.1103/PhysRevB.90.161106
24. J. Sýkora, T. Holder, W. Metzner, Phys. Rev. B **97**, 155159 (2018). https://doi.org/10.1103/PhysRevB.97.155159. https://link.aps.org/doi/10.1103/PhysRevB.97.155159
25. T. Senthil, R. Shankar, Phys. Rev. Lett. **102**, 046406 (2009). https://doi.org/10.1103/PhysRevLett.102.046406
26. I. Mandal, Phys. Rev. Research **2**, 043277 (2020). https://doi.org/10.1103/PhysRevResearch.2.043277. https://link.aps.org/doi/10.1103/PhysRevResearch.2.043277
27. I. Mandal, R.M. Fernandes, Phys. Rev. B **107**, 125142 (2023). https://doi.org/10.1103/PhysRevB.107.125142. https://link.aps.org/doi/10.1103/PhysRevB.107.125142
28. P. Rao, F. Piazza, Phys. Rev. Lett. **130**, 083603 (2023). https://doi.org/10.1103/PhysRevLett.130.083603. https://link.aps.org/doi/10.1103/PhysRevLett.130.083603

# Chapter 6
# Non-Fermi Liquids by Critical Cavity Photons at the Onset of Superradiance

## 6.1 Introduction

Cavity-confined photons have emerged as a powerful and versatile tool for engineering strong electron-electron interactions via light-matter coupling [1–7], motivating us to explore the possibility of non-Fermi liquid (NFL) phases arising in cavity quantum electrodynamics (QED) set-ups involving two-dimensional (2d) crystalline lattices [8–16]. The rapidly advancing field of cavity QED offers the prospect of realizing strong coupling between cavity photons and fermionic matter, from 2d layered heterostructures [17] to synthetic ultracold atomic arrays [8–10]. Although the finite spatial extent of the cavity endows the photonic modes with an effective mass, the system can be driven toward a superradiant phase in which the cavity photons become massless at a continuous phase transition [11–16]. In the superradiant phase, the ground state hosts a macroscopic occupation of coherent photons—effectively, a photon condensate—coupling the lattice atoms to a spatially uniform, single-mode electromagnetic field. Accessing this transition demands an exceptionally strong atom-field coupling. Superradiant phase transitions have been realized experimentally across a range of platforms, including optically pumped gases [18], photoexcited semiconducting quantum dots [19, 20], and pumped ultracold gases confined in ultrahigh-finesse optical cavities [21]. It is important to note, however, that all these experiments rely on an external drive, and a true equilibrium realization of the superradiant transition remains elusive. We proceed nonetheless with the expectation that such an equilibrium set-up will be achieved in the near future.

## 6.2 Model

Our starting point is the non-interacting tight-binding Hamiltonian on the honeycomb lattice, the paradigmatic model for graphene, which has been studied extensively in the literature [22]. At half-filling, Dirac cones emerge at two types of inequivalent

 
I. Mandal, *Controlled Description Of Non-Fermi Liquids Using Field-Theoretic Methods*, SpringerBriefs in Physics, https://doi.org/10.1007/978-3-032-21618-2_6

valleys, labelled as $K_\varsigma$ (with $\varsigma = \pm$), and the two types of Dirac points are exchanged under time-reversal symmetry. We focus on the low-energy dynamics of fermions with momenta close to $K_\varsigma$, which is captured by taking the continuum limit of the tight-binding model. Upon doping away from the Dirac points, the system evolves into a Fermi liquid with a nonparabolic dispersion and singly-connected convex Fermi surfaces (FSs). At sufficiently low doping, the dispersion remains linear and isotropic, placing the system in the regime of the so-called Dirac Fermi liquid (DFL) [23].

The band structure is obtained by expanding the momentum $\mathbf{P}$ around the valley $K_\varsigma$ as $\mathbf{P} = \mathbf{K}_\varsigma + \mathbf{p}$, where $|\mathbf{p}|/|\mathbf{K}_\varsigma| \ll 1$. At leading order in this small parameter, one recovers the linear dispersion characteristic of the DFL. Retaining corrections up to order $(|\mathbf{p}|/|\mathbf{K}_\varsigma|)^2$ introduces trigonal-warping of the bands [22, 24, 25] in the vicinity of $K_\varsigma$. Working in a diagonal electron-hole basis and suppressing the spin degrees of freedom, the noninteracting Hamiltonian takes the form [23],

$$H_0 = \sum_{\varsigma,\mathbf{p}} \left[ \left(\epsilon_{\varsigma,\mathbf{p},+} - \mu_c\right) \alpha^\dagger_{\varsigma,\mathbf{p}} \alpha_{\varsigma,\mathbf{p}} + \left(\epsilon_{\varsigma,\mathbf{p},-} - \mu_c\right) \beta^\dagger_{\varsigma,\mathbf{p}} \beta_{\varsigma,\mathbf{p}} \right], \tag{6.1}$$

where $\mu_c$ is the chemical potential. The dispersion, accounting for trigonal-warping corrections, reads

$$\begin{aligned} \epsilon_{\varsigma,\mathbf{p},\lambda}(\theta_p) &= \epsilon^{\mathrm{D}}_{\mathbf{p},\lambda} + \epsilon^{\mathrm{TW}}_{\varsigma,\mathbf{p},\lambda}(\theta_p), \quad \epsilon^{\mathrm{D}}_{\mathbf{p},\lambda} = \lambda\, v_{\mathrm{D}}\, |\mathbf{p}|, \\ \epsilon^{\mathrm{TW}}_{\varsigma,\mathbf{p},\lambda}(\theta_p) &= \lambda\, \varsigma\, \frac{v_{\mathrm{D}}\, |\mathbf{p}|^2}{2\tilde{\rho}} \cos(3\theta_p), \quad \theta_p = \arctan\left(\frac{p_y}{p_x}\right), \end{aligned} \tag{6.2}$$

where $\alpha^\dagger_{\varsigma,\mathbf{p}}$ ($\beta^\dagger_{\varsigma,\mathbf{p}}$) creates an electron (hole) in the conduction (valence) band in the vicinity of valley $K_\varsigma$. The parameter $\lambda = \pm$ labels the band, $v_{\mathrm{D}}$ denotes the Fermi velocity at the Dirac point, and $\tilde{\rho}$ encodes the strength of trigonal-warping, scaling inversely with the nearest-neighbor hopping distance. Our coordinate choice ensures that the Dirac and warping terms enter with the same sign for the valley $K_+$. Provided the warping remains sufficiently weak, both FSs retain global convexity, as depicted in Fig. 6.1.

## 6.2.1 *Location of Hot-Spots Paired by CDW Ordering*

The electron-photon coupling is implemented through the Peierls substitution, which ties the fermion-boson vertex to the gradient of the Hamiltonian with respect to the vector potential [16]. For the Hamiltonian in Eq. (6.1), the current operator is extracted via the standard formula $\mathbf{j} = -\frac{\delta}{\delta \mathbf{A}} H_0(\mathbf{p} - e\mathbf{A})$, where $\mathbf{A}$ denotes the vector potential. The completely filled valence band contributes no net current, so we focus exclusively on $\lambda = +$, yielding the intraband current

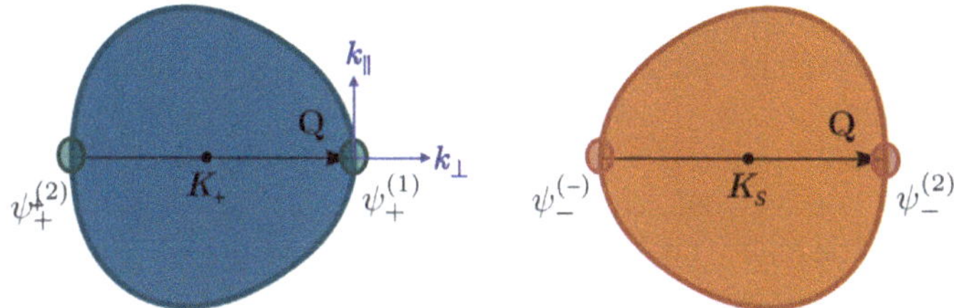


**Fig. 6.1** Schematic illustration of the two trigonally-warped Fermi surfaces situated at the neighboring valleys, $K_+$ and $K_-$. The hot-spot pairs on each surface, linked by the incommensurate wavevector $\mathcal{Q}$ (depicted by the black arrow), are emphasised. In the $K_+$ ($K_-$) valley, the fermionic degrees of freedom localised near the right and left hot-spots are denoted $\psi_+^{(1)}$ ($\psi_-^{(2)}$) and $\psi_+^{(2)}$ ($\psi_-^{(1)}$), respectively. These fermions interact with the critical cavity-photon modes, which carry momenta centered about $\mathcal{Q}$

$$\mathbf{j} = \sum_{\varsigma,\mathbf{p}} \mathbf{v}_{\varsigma,\mathbf{p}} \alpha^\dagger_{\varsigma,\mathbf{p}} \alpha_{\varsigma,\mathbf{p}}, \quad \mathbf{v}_{\varsigma,\mathbf{p}} = \nabla_{\mathbf{k}} \epsilon_{\varsigma,\mathbf{p},+}. \tag{6.3}$$

For brevity, we drop the spin label throughout what follows. An important consequence is that the polarization function—which controls the possible appearance of a CDW gap—arises from the current-current correlator and therefore involves no intervalley scattering: CDW bosons can only link hot-spots belonging to the same valley.

The derivation of the polarization function is a standard exercise well covered in existing literature and serves to establish the conditions under which a CDW gap opens. The mechanism relies on an enhancement of the low-energy phase space when electrons scatter between hot-spots whose tangent vectors are either parallel or antiparallel. The setup bears a close resemblance to the current-current correlator analysis of Ref. [23], which treated doped monolayer graphene with an effective four-fermion interaction generated by the bare Coulomb potential, also incorporating trigonal-warping. The crucial difference is that our model contains no four-fermion interaction, thereby eliminating intervalley scattering processes altogether. Had such a term been present, CDW instabilities with wavevector $\mathcal{Q}$ linking time-reversed patches on the two distinct FSs centered at $K_+$ and $K_-$ would have been accessible, as explored in Ref. [26]. We further remark that the supplemental calculations in Ref. [16] assume circular FSs at each valley, with hot-spot pairs sharing the same curvature. In our case, the FS are noncircular due to trigonal-warping, and the hot-spots connected by $\mathcal{Q}$ typically possess different curvatures.

### 6.2.2 *Patch Theory Using Time-Reversed Partners*

To construct a patch theory, we introduce two-component spinors formed from time-reversed partners [27, 28], which in this particular setting case effectively correspond to fermionic operators on the conjugate valleys $K_+$ and $K_-$. Each valley supports three distinct pairs of hot-spots, with the members of each pair connected by a CDW boson. These three pairs are related to one another by $2\pi/3$ rotations, reflecting the threefold rotational symmetry inherent to the honeycomb lattice. The momenta of the three CDW bosons, centered at wavevectors $\mathcal{Q}$, are similarly related by $2\pi/3$ rotations. Since the cavity photons carry no self-interactions, the theory factorizes into three decoupled sectors, each containing two pairs of hot-spots with parallel or antiparallel tangent vectors, coupled to a single CDW boson species. It therefore suffices to study a single such sector, to which we restrict our attention throughout the remainder of this chapter. Working in patch coordinates [27–32], the effective action for the patches located at $\theta_p = 0$ and $\theta_p = \pi$ reads [33],

$$
\begin{aligned}
S = & \sum_{\substack{\varsigma=\pm \\ n=1,2}} \int_k \{\psi_\varsigma^{(n)}(k)\}^\dagger \left[-i\, k_0 - (-1)^n \,\varsigma\, v_F^{(n)}\, k_1 + \kappa^{(n)}\, k_2^2\right] \psi_\varsigma^{(n)}(k) \\
& + \frac{1}{2}\int_k \phi(k)\left(k_0^2 + k_1^2 + k_2^2\right)\phi(-k) \\
& + \frac{e}{2}\int_k\int_q \left[\phi(q)\left\{\psi_+^{(1)}(k+q)\right\}^\dagger \psi_+^{(2)}(k) + \phi(-q)\left\{\psi_+^{(2)}(k-q)\right\}^\dagger \psi_+^{(1)}(k)\right] \\
& + \frac{e}{2}\int_k\int_q \left[\phi(q)\left\{\psi_-^{(2)}(k+q)\right\}^\dagger \psi_-^{(1)}(k) + \phi(-q)\left\{\psi_-^{(1)}(k-q)\right\}^\dagger \psi_-^{(2)}(k)\right].
\end{aligned}
\tag{6.4}
$$

The three-vector $k = (k_0, \boldsymbol{k})$ comprises the Matsubara frequency, $k_0$, and the spatial momentum, $\boldsymbol{k} = (k_1, k_2) \equiv (k_\perp, k_\parallel)$. The integrals are denoted via the shorthand $\int_k \equiv \int dk_0\, d^d\boldsymbol{k}/(2\pi)^{d+1}$ and $d = 2$ spatial dimensions. In the neighborhood of $K_\varsigma$, the low-energy fermionic excitations localized near the two hot-spots are labeled $\psi_\varsigma^{(1)}(k)$ and $\psi_\varsigma^{(2)}(k)$, as illustrated in Fig. 6.1. The CDW bosonic field $\phi(k)$, arising from the cavity photons, carries frequency $k_0$ and momentum $\mathcal{Q} + \boldsymbol{k}$. Precisely at the superradiant quantum critical point, these cavity bosons lose their mass, as manifest in the purely quadratic bosonic sector of the action. We rescale fermionic momenta such that for the fields $\psi_+^{(1)}(k)$ and $\psi_-^{(1)}(k)$, both the Fermi velocity and curvature are set to unity: $v_F^{(1)} = \kappa^{(1)} = 1$. For the remaining hot-spots, we adopt the notation $v_F^{(2)} = v$ and $\kappa^{(2)} = \kappa$. The curvature $\kappa$ can become negative when the corresponding hot-spot is concave. Although the bare bosonic velocity differs in general from its fermionic counterpart, we normalize it to unity, as it does not enter the infrared effective theory: near the quantum critical point, the bosonic propagator is governed entirely by particle-hole fluctuations of the FS at low energies.

Before rescaling, the Fermi velocity at angular position $\theta_p$ reads $v_{F,\varsigma}(\theta_p) = v_D\left[1 + \varsigma\cos(3\theta_p)/\rho\right]$, with $\rho \simeq \tilde{\rho}\, v_D/\mu_c$, and we write $\kappa_\varsigma(\theta_p)$ for half the curvature. Upon evaluation at $\theta_p = 0$ and $\theta_p = \pi$ (cf. Fig. 6.1), these become $v_F^{(n)}$ and $\kappa^{(n)}$. Exploiting the freedom to rescale, we pass to the dimensionless ratios $v_{F,\varsigma}(\theta_p) \to v_{F,\varsigma}(\theta_p)/v_{F,+}(0)$ and $\kappa_\varsigma(\theta_p) \to \kappa_\varsigma(\theta_p)/\kappa_+(0)$, as depicted in Fig. 6.2 for the valley

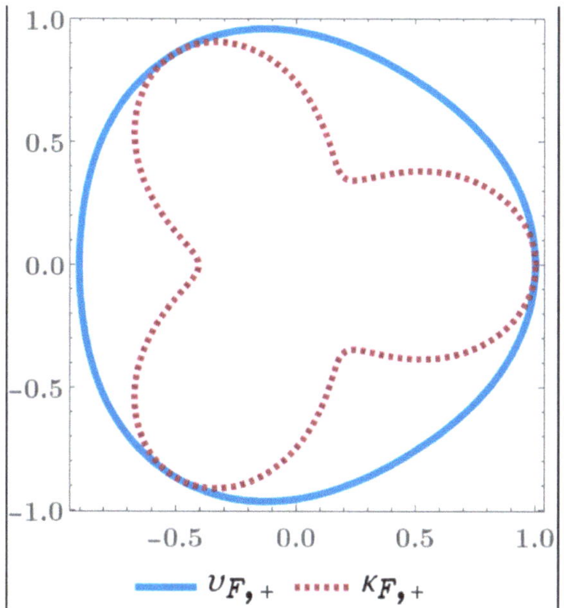


**Fig. 6.2** Representative parameters characterising the FS at valley $K_+$, shown for a configuration in which all hot-spots exhibit positive curvature

$K_+$. This choice renders $v_F^{(1)} \equiv v_{F,+}(0)$ and $\kappa^{(1)} \equiv \kappa_+(0)$ both equal to unity—these correspond to the right-hand hot-spot of $K_+$ and the left-hand hot-spot of $K_-$—while the conjugate hot-spots carry $v_F^{(2)} \equiv v = v_{F,+}(\pi)$ and $\kappa^{(2)} \equiv \kappa = \kappa_+(\pi)$, subject to the global convexity requirement $1 - \kappa/v \geq 0$. Throughout, we work in parameter regimes where both FSs remain everywhere convex.

Within the patch-coordinate framework, essential for capturing the correct energy-windows of the fermions and bosons, the key scaling-dimension assignments are: $[\mathbf{K}] = 1$ and $[k_d] = 1/2$. Examining the kinetic terms in the action then reveals the engineering dimensions of $[\psi_{\varsigma}^{(n)}] = [\phi_\pm] = -7/4$. Substituting these values into the interaction vertex yields $[e] = 1/4$. This establishes $e$ as a relevant operator and presages the emergence of a NFL, in direct analogy with the systems analyzed in Refs. [27–29, 32]. To gain analytical control, we invoke dimensional regularization, continuously varying the co-dimension of the FS as an auxiliary mathematical tool. This procedure identifies the upper critical dimension $d = d_c$, defined as the point at which the one-loop fermionic self-energy acquires a logarithmic divergence in the Wilsonian cutoff $\Lambda$, separating the NFL regime from marginal-Fermi-liquid behaviour.

Ensuring that the theory remains analytic in momentum space (a requirement equivalent to preserving locality in position space), when the co-dimension is extended to generic values, necessitates the introduction of two-component spinors as follows [27, 28, 33]

$$\Psi_1^T(k) = \begin{bmatrix} \psi_+^{(1)}(k) & \left\{\psi_-^{(1)}\right\}^\dagger(-k) \end{bmatrix} \text{ and } \Psi_2^T(k) = \begin{bmatrix} \psi_-^{(2)}(k) & \left\{\psi_+^{(2)}\right\}^\dagger(-k) \end{bmatrix}, \tag{6.5}$$

together with their conjugates,

$$\bar{\Psi}_n \equiv \Psi_n^\dagger \gamma_0 \quad \text{for } n \in \{1, 2\}. \tag{6.6}$$

With these spinors in hand, we construct an effective action describing the patches of the one-dimensional FS near the hot-spots, now embedded in a $d$-dimensional momentum space [27–29, 31, 32]. The resulting low-energy effective action reads,

$$\begin{aligned} S &= \sum_n \int_k \bar{\Psi}_n(k)\, i\left[\Gamma \cdot \mathbf{K} + \gamma_{d-1}\, \delta_k^{(n)}\right] \Psi_n(k) + \frac{1}{2}\int_k k_d^2\, \phi(k)\, \phi(-k) \\ &\quad - \left[\frac{i\, e\, \mu^{x/2}}{2} \int_k \int_q \phi(q)\, \bar{\Psi}_1(k+q)\, \Psi_2(-k) + \text{h.c.}\right], \\ x &= \frac{5}{2} - d, \quad \delta_k^{(1)} = k_{d-1} + k_d^2, \quad \delta_k^{(2)} = \upsilon\, k_{d-1} + \kappa\, k_d^2. \end{aligned} \tag{6.7}$$

The $(d-1)$-dimensional vector $\mathbf{K} \equiv (k_0, k_1, \ldots, k_{d-2})$ assembles the frequency and the $(d-2)$ momentum components introduced by extending the co-dimension. The original momentum components along the $k_1$- and $k_2$-directions are relabelled as $k_{d-1}$ and $k_d$, respectively. Adopting these notations implies that, in the $d$-dimensional momentum space, $\{k_1, \ldots, k_{d-1}\}$ spans the $(d-1)$ directions perpendicular to the FS, while $k_d$ runs parallel to it. The symbol $\mathbf{\Gamma} \equiv (\gamma_0, \gamma_1, \ldots, \gamma_{d-2})$ denotes a $(d-1)$-dimensional vector of matrices whose contraction with $\mathbf{K}$ enters the fermionic kinetic term. For our purpose, it suffices to work with $2 \times 2$ Pauli matrices, taking $\gamma_0 = \sigma_y$ and $\gamma_{d-1} = \sigma_x$, since our ultimate goal is to analytically continue to $d = 2$, which is the physical spatial-dimensionality of the system. A floating mass scale, $\mu \sim \Lambda$, raised to the power $x/2$, is introduced to render the coupling constant $e$ dimensionless.

The kinetic parts of the action in Eq. (6.7) remain invariant under the following scaling transformations:

$$\begin{aligned} &\mathbf{K} = \frac{\mathbf{K}'}{b}, \quad k_{d-1} = \frac{k'_{d-1}}{b}, \quad k_d = \frac{k'_d}{\sqrt{b}}, \\ &\Psi_n(k) = b^{\frac{2d+3}{4}}\, \Psi'_n(k'), \quad \phi(k) = b^{\frac{2d+3}{4}}\, \phi'(k'). \end{aligned} \tag{6.8}$$

This follows from the fact that $[\mathbf{K}] = 1$ and $[k_d] = 1/2$, which are the defining characteristics of the patch coordinates. We note that the fermions near the two hot-spots, which interact strongly with the bosons, satisfy $|k_d| \gg k_{d-1}$, since any scattering event away from the FS incurs a large energy-cost. In the bosonic kinetic term, only the contribution proportional to $k_d^2$ is retained, as the part involving $(\mathbf{K}^2 + k_{d-1}^2)$ is irrelevant under the scaling-relations written above.

From Eq. (6.7), the bare fermionic propagator is deduced to be

$$G_n(k) \equiv \left\langle \Psi_n(k)\, \bar{\Psi}_n(k) \right\rangle_0 = -\,i\, \frac{\boldsymbol{\Gamma} \cdot \mathbf{K} + \gamma_{d-1}\, \delta_k^{(n)}}{\mathbf{K}^2 + \delta_k^2}. \tag{6.9}$$

Similarly, the bare bosonic propagator is given by $D_{(0)}(k) = 1/k_d^2$.

The parameter $x$ establishes that the marginality of the coupling constant, $e$, occurs at the upper critical dimension, $d_c = 5/2$. It also tells us that $e$ behaves as a relevant (irrelevant) operator for $d < 5/2$ ($d > 5/2$). The strategy is to construct a controlled perturbative description of the interacting phase by working in $d = 5/2 - \epsilon$ dimensions, treating $\epsilon$ as an expansion parameter. The physical theory is then accessed by setting $\epsilon = 1/2$ upon the implementation of a systematic order-by-order expansion in $\epsilon$.

## 6.3 One-Loop Feynman Diagrams

We now present the results for all Feynman diagrams contributing at one-loop order, albeit with the loop-ordering for the fermionic self-energy and vertex corrections being dictated by considering the dressed bosonic propagator. This amounts to rearranging the perturbative expansion such that the one-loop bosonic self-energy is already included at the 'zero'-th order.

### 6.3.1 One-Loop Boson Self-Energy

We begin by computing the one-loop bosonic self-energy, which starts from the expression,

$$\begin{aligned}
\Pi_1(q) &= -\frac{\left(i\, e\, \mu^{\frac{x}{2}}\right)^2}{2} \int_k \mathrm{Tr}\left[G_1(k+q)\, G_2(k)\right] \\
&= e^2\, \mu^x \int_k \frac{\mathbf{K} \cdot (\mathbf{K} + \mathbf{Q}) + \delta_k^{(1)}\, \delta_{k+q}^{(2)}}{\left[\mathbf{K}^2 + \delta_k^2\right]\left[(\mathbf{K} + \mathbf{Q})^2 + \delta_{k+q}^2\right]} \\
&= e^2\, \mu^x \int_k \frac{\mathbf{K} \cdot (\mathbf{K} + \mathbf{Q}) + \upsilon\, \delta_k^{(1)} \left\{k_{d-1} + q_{d-1} + \frac{\kappa}{\upsilon}\, (k_d + q_d)^2\right\}}{\upsilon^2\left[\mathbf{K}^2 + \delta_k^2\right]\left[\frac{(\mathbf{K}+\mathbf{Q})^2}{\upsilon^2} + \left\{k_{d-1} + q_{d-1} + \frac{\kappa}{\upsilon}\, (k_d + q_d)^2\right\}^2\right]}.
\end{aligned} \tag{6.10}$$

Proceeding with the integration over $k_{d-1}$ then yields [33]

$$\Pi_1(q) = \frac{e^2\,\mu^x}{2} \int \frac{dk_d\, d\mathbf{K}}{(2\pi)^d} \frac{\left(\mathbf{K} + \frac{|\mathbf{K}+\mathbf{Q}|}{\upsilon}\right)\left[\mathbf{K}\cdot(\mathbf{K}+\mathbf{Q}) + \mathbf{K}\,\frac{|\mathbf{K}+\mathbf{Q}|}{\upsilon}\right]}{\upsilon^2\,\mathbf{K}\,\frac{|\mathbf{K}+\mathbf{Q}|}{\upsilon}\left[\left\{q_{d-1} + \frac{\kappa}{\upsilon}\,(k_d+q_d)^2 - k_d^2\right\}^2 + \left(\mathbf{K} + \frac{|\mathbf{K}+\mathbf{Q}|}{\upsilon}\right)^2\right]}$$
$$= \frac{e^2\,\mu^x}{2} \int \frac{dk_d\, d\mathbf{K}}{(2\pi)^d} \frac{\left(\mathbf{K} + \frac{|\mathbf{K}+\mathbf{Q}|}{\upsilon}\right)\left[\mathbf{K}\cdot(\mathbf{K}+\mathbf{Q}) + \mathbf{K}\,\frac{|\mathbf{K}+\mathbf{Q}|}{\upsilon}\right]}{\upsilon\,\mathbf{K}\,|\mathbf{K}+\mathbf{Q}|\left[\Upsilon^2(k,q) + \left(\mathbf{K} + \frac{|\mathbf{K}+\mathbf{Q}|}{\upsilon}\right)^2\right]}, \tag{6.11}$$

where

$$\Upsilon(q,k) = \begin{cases} \left(1-\frac{\kappa}{\upsilon}\right)\left[k_d^2 - \frac{\upsilon\, e_q}{\upsilon-\kappa}\right] & \text{for } \upsilon \neq \kappa \\ \delta_q^{(1)} + 2\,k_d\,q_d & \text{for } \upsilon = \kappa \end{cases}, \tag{6.12}$$

and

$$e_q = \frac{\kappa\, q_d^2}{\upsilon-\kappa} + q_{d-1}. \tag{6.13}$$

In the special case where $\upsilon = \kappa$, this reduces to

$$\Pi_1(q) = \frac{e^2\,\mu^x}{4} \int \frac{d\mathbf{K}}{(2\pi)^d} \int_{-\infty}^{\infty} dk_d \frac{\left(\mathbf{K} + \frac{|\mathbf{K}+\mathbf{Q}|}{\upsilon}\right)\left[\mathbf{K}\cdot(\mathbf{K}+\mathbf{Q}) + \mathbf{K}\,|\mathbf{K}+\mathbf{Q}|\right]}{\sqrt{|q_d|}\,\upsilon\,\mathbf{K}\,|\mathbf{K}+\mathbf{Q}|\left[k_d^2 + \left(\mathbf{K} + \frac{|\mathbf{K}+\mathbf{Q}|}{\upsilon}\right)^2\right]}$$
$$= \frac{e^2\,\mu^x}{8\,|q_d|\,\upsilon}\, I_1(d,\mathbf{Q}), \tag{6.14}$$

with

$$I_1(d,\mathbf{Q}) = \int \frac{d^{d-1}\mathbf{K}}{(2\pi)^{d-1}} \left[\frac{\upsilon\,\mathbf{K}\cdot(\mathbf{K}+\mathbf{Q})}{\mathbf{K}\,|\mathbf{K}+\mathbf{Q}|} + 1\right]. \tag{6.15}$$

When $\upsilon \neq \kappa$, we perform the change of variables, $u = k_d^2$, which yields a Jacobian factor $1/(2\sqrt{u}) = 1/(2|k_d|)$. Inspecting the denominator of the second factor in the integrand reveals that the dominant contribution is concentrated near $u \sim e_q$ in the regime $|\mathbf{Q}| \ll \kappa\, q_d^2$. Treating $e_q$ as positive and noting that the typical energy scales enforce $q_d \gg q_{d-1}$, we replace $|k_d|$ in the Jacobian by $\sqrt{\upsilon\, e_q/(\upsilon-\kappa)}$. The integral then evaluates to [33]

$$\Pi_1(q) = -\,\beta_d\, e^2\, \mu^x\, \frac{|\boldsymbol{Q}|^{d-1}}{f(q)}, \quad f(q) = \begin{cases} \sqrt{\upsilon\,(\upsilon-\kappa)}\,\sqrt{e_q}\,\Theta(e_q) & \text{for } \upsilon \neq \kappa \\ 2\,|q_d| & \text{for } \upsilon = \kappa \end{cases},$$

$$\beta_d = \frac{\Gamma^2\left(\frac{d}{2}\right)}{2^d\,\pi^{\frac{d-1}{2}}\,\left|\cos\left(\frac{\pi d}{2}\right)\right|\,\Gamma\left(\frac{d-1}{2}\right)\Gamma(d)}. \tag{6.16}$$

The bare bosonic propagator, $D_{(0)}(k)$, lacks any $\mathbf{K}$-dependence, rendering loop integrals involving it divergent, unless one performs a resummation that generates a dynamical dispersion along these directions. We remedy this by dressing the propagator with the lowest-order finite correction $\Pi_1(k)$ from the one-loop bosonic self-energy, which scales as $|\mathbf{K}|^{d-1}/f(k)$, and incorporate this correction into all subsequent loop computations. Operationally, this means adopting the dressed propagator $D_{(1)}(k) = \left[\left(D_{(0)}(k)\right)^{-1} - \Pi_1(k)\right]^{-1}$, which effectively reorganizes the perturbative series by promoting the $\mathbf{K}$-dependent finite piece of the one-loop bosonic self-energy to zeroth order. The correction $\Pi_1(k)$ is none other than the well-known *Landau-damping* term, responsible for inducing the distinctive $\text{sgn}(k_0)|k_0|^{2/3}$ frequency scaling in the fermionic self-energy—a universal hallmark of NFL physics at quantum critical points in diverse strongly correlated systems [27–29, 31, 34–36].

### 6.3.2 One-Loop Fermion Self-Energies

Two distinct one-loop diagrams contribute to the fermionic self-energy, which we compute separately below. Incorporating the dressed bosonic propagator, the corresponding starting expressions are:

$$\begin{aligned}\Sigma_1(q) &= \frac{\left(i\,e\,\mu^{\frac{x}{2}}\right)^2}{2} \int_k G_2(q-k)\,D_{(1)}(k) \\ &= \frac{i\,e^2\,\mu^x}{2} \int_k \frac{1}{k_d^2 + \beta_d\, e^2\,\mu^x\,\frac{|\mathbf{K}|^{d-1}}{f(k)}}\,\frac{\gamma_{d-1}\,\delta^{(2)}_{q-k} + \Gamma\cdot(\boldsymbol{Q}-\mathbf{K})}{(\boldsymbol{Q}-\mathbf{K})^2 + \left[\delta^{(2)}_{q-k}\right]^2},\end{aligned} \tag{6.17}$$

and

$$\begin{aligned}\Sigma_2(q) &= \frac{\left(i\,e\,\mu^{\frac{x}{2}}\right)^2}{2} \int_k G_1(q-k)\,D_{(1)}(k) \\ &= \frac{i\,e^2\,\mu^x}{2} \int_k \frac{1}{k_d^2 + \beta_d\, e^2\,\mu^x\,\frac{|\mathbf{K}|^{d-1}}{f(k)}}\,\frac{\gamma_{d-1}\,\delta^{(1)}_{q-k} + \Gamma\cdot(\boldsymbol{Q}-\mathbf{K})}{(\boldsymbol{Q}-\mathbf{K})^2 + \left[\delta^{(1)}_{q-k}\right]^2},\end{aligned} \tag{6.18}$$

where

$$\begin{aligned}\delta^{(2)}_{q-k} &= -\,\upsilon\, k_{d-1} + \delta^{(2)}_{q} + \kappa\left(k_d^2 - 2\, q_d\, k_d\right)\\ &= \upsilon\left[q_{d-1} + \frac{\kappa\, q_d^2}{\upsilon} + \frac{\kappa}{\upsilon}\left(k_d^2 - 2\, q_d\, k_d\right) - k_{d-1}\right].\end{aligned} \tag{6.19}$$

Evaluating at $d = d_c - \epsilon$, the singular components of the self-energy can be extracted as follows:

1. For $\upsilon \neq \kappa$:

$$\begin{aligned}\Sigma_1(q) &= -\,\frac{\mathcal{U}_1\, e^{\frac{4}{3}}}{\epsilon}\,\frac{\left[\kappa\,(2\,\upsilon - \kappa)\right]^{\frac{1}{6}}}{\upsilon}\, i\,(\Gamma\cdot \boldsymbol{Q}) + \mathcal{O}\left(\epsilon^0\right),\\ \Sigma_2(q) &= -\,\frac{\mathcal{U}_1\, e^{\frac{4}{3}}}{\upsilon^{2/3}\,\epsilon}\, i\,(\Gamma\cdot \boldsymbol{Q}) + \mathcal{O}\left(\epsilon^0\right),\quad \mathcal{U}_1 = \frac{\left[\Gamma\left(\frac{1}{4}\right)\Gamma\left(\frac{5}{4}\right)\right]^{1/3}}{6\times 3^{1/6}\,\pi^{4/3}}.\end{aligned} \tag{6.20}$$

2. For $\upsilon = \kappa$:

$$\Sigma_2(q) = \upsilon\, \Sigma_1(q),\quad \Sigma_1(q) = -\,\frac{\mathcal{U}_2\, e^{\frac{4}{3}}}{\upsilon\,\epsilon}\, i\,(\Gamma\cdot \boldsymbol{Q}) + \mathcal{O}\left(\epsilon^0\right),\quad \mathcal{U}_2 = \frac{2^{1/3}}{3^{7/6}}. \tag{6.21}$$

In this expression, the logarithmic divergences that would emerge from integrating out high-energy modes in a Wilsonian approach manifest themselves as simple poles at $\epsilon = 0$ (within our dimensional regularization framework).

### 6.3.3 One-Loop Vertex-Corrections

The one-loop fermion-boson vertex functions, $\Gamma_{12}(q, p)$ and $\Gamma_{21}(q, p)$, are generically functions of the two external frequency-momentum variables, $p$ and $q$. For the purpose of extracting the leading-order singular behavior proportional to $1/\epsilon$, we need only evaluate these vertices in the limit $p \to 0$, whereupon the corresponding loop-integrals reduce to

$$\begin{aligned}\Gamma_{n_1 n_2}(q, 0) &= \frac{e^2\,\mu^x}{2}\int_k G_{n_1}(k)\, G_{n_2}(k)\, D_{(1)}(k-q)\\ &= \frac{e^2\,\mu^x}{2}\int_k D_{(1)}(k-q)\,\frac{\delta_k^{(n_1)}\,\delta_k^{(n_2)} + \mathbf{K}^2 - \gamma_{d-1}\,(\boldsymbol{\Gamma}\cdot\mathbf{K})\left[\delta_k^{(n_1)} + \delta_k^{(n_2)}\right]}{\left[\mathbf{K}^2 + \left\{\delta_k^{(n_1)}\right\}^2\right]\left[\mathbf{K}^2 + \left\{\delta_k^{(n_2)}\right\}^2\right]}.\end{aligned} \tag{6.22}$$

Since $\Gamma_{12}(q,0) = \Gamma_{21}(q,0)$, it suffices to evaluate

$$\Gamma_{12}(q,0) = \frac{e^2\,\mu^x}{2}\int_k \frac{\delta_k^{(1)}\,\delta_k^{(2)} + \mathbf{K}^2 - \gamma_{d-1}\,(\boldsymbol{\Gamma}\cdot\mathbf{K})\left[\delta_k^{(1)} + \delta_k^{(2)}\right]}{\left[\mathbf{K}^2 + \left\{\delta_k^{(1)}\right\}^2\right]\left[\mathbf{K}^2 + \left\{\delta_k^{(2)}\right\}^2\right]}. \tag{6.23}$$

The integral vanishes identically in the special case $\upsilon = \kappa$, prompting us to concentrate on the generic situation where $\upsilon \neq \kappa$. In both scenarios, however, the outcome is the same: no singular contribution emerges.

## 6.4 RG Flows for $\upsilon \neq \kappa$

In our QFT treatment, the action appearing in Eq. (6.7) is designated as the *physical action*, formulated at an energy scale, $\mu \sim \Lambda$, comprising non-divergent physically-observable quantities. Loop corrections, however, generate contributions that exhibit either logarithmic or power-law divergences in $\Lambda$. We regulate these ultraviolet singularities through renormalization, implemented via dimensional regularization, a procedure in which divergences re-emerge as poles in the parameter $\epsilon$ in the limit $\epsilon \to 0$. Our computations have been carried out at one-loop order. The calculations are conducted within the minimal subtraction (MS) renormalization scheme [37, 38], wherein divergent contributions from loop diagrams are systematically removed through the introduction of counterterms. We specifically employ the modified minimal subtraction ($\overline{\text{MS}}$) variant, which absorbs not only the $1/\epsilon$ pole itself, but also the universal finite piece proportional to $\epsilon^0$ that accompanies it.

The *counterterm action*, constructed to absorb the singular contributions arising from quantum corrections, mirrors the structure of the physical action in Eq. (6.7) and takes the explicit form,

$$\begin{aligned}
S_{CT} &= \int_k \bar{\Psi}_1(k)\, i\left[A_1\,\Gamma\cdot\mathbf{K} + \gamma_{d-1}\left(A_2\,k_{d-1} + A_3\,k_d^2\right)\right]\Psi_1(k) \\
&+ \int_k \bar{\Psi}_2(k)\, i\left[A_4\,\Gamma\cdot\mathbf{K} + \gamma_{d-1}\left(A_5\,\upsilon\,k_{d-1} + A_6\,\kappa\,k_d^2\right)\right]\Psi_2(k) \\
&+ \frac{1}{2}\int_k A_7\,k_d^2\,\phi(k)\,\phi(-k) - \left[\frac{i\,e\,\mu^{x/2}}{2}\int_k\int_q A_8\,\phi(q)\,\bar{\Psi}_1(k+q)\,\Psi_2(-k) + \text{h.c.}\right].
\end{aligned} \tag{6.24}$$

The coefficients appearing in the counterterms are given by the power series,

$$A_\zeta = \sum_{n=1}^{\infty}\frac{Z_\zeta^{(n)}}{\epsilon^n} \quad \text{with} \quad \zeta \in [1,8], \tag{6.25}$$

chosen so as to cancel the divergent contributions $\propto 1/\epsilon^n$ arising from the loop-level Feynman diagrams. The $(d-1)$-dimensional rotational invariance in the space perpendicular to the FS ensures that each term in $\mathbf{\Gamma}\cdot\mathbf{K}$ is renormalized identically.

With these elements in hand, we formally subtract $S_{CT}$ from the so-called *bare* action,

$$\begin{aligned} S_{\text{bare}} =& \sum_n \int_{k^B} \bar{\Psi}_n^B(k^B)\, i\left[\mathbf{\Gamma}\cdot\mathbf{K}^B + \gamma_{d-1}\,\delta_{k^B}^{(n)}\right]\Psi_n^B(k^B) + \int_{k^B} \frac{(k_d^B)^2\,\phi^B(k^B)\,\phi^B(-k^B)}{2} \\ & - \left[\frac{i\,e^B}{2}\int_{k^B}\int_{q^B} \phi^B(q^B)\,\bar{\Psi}_1^B(k^B+q^B)\,\gamma_{d-1}\,\Psi_2^B(-k^B) + \text{h.c.}\right]. \end{aligned} \tag{6.26}$$

By construction, the resulting *physical* effective action $S$ contains only well-defined, non-divergent quantum parameters, whereas $S_{\text{bare}}$ is formulated in terms of *bare quantities* that may diverge. Throughout, the superscript "$B$" designates bare fields, couplings, frequencies, and momenta. This framework enables physical observables to be extracted from renormalized coupling constants, whose evolution is governed by RG flow equations—differential equations that track how the couplings vary as the floating energy scale $\mu\, e^{-l}$ changes, or equivalently, as the logarithmic length scale $l$ increases.

The RG flow equations are derived by connecting the bare quantities to their renormalized counterparts (those without the superscript "$B$") through multiplicative $Z_\zeta$-factors:

$$S_{\text{bare}} = S + S_{CT}\,, \quad Z_\zeta = 1 + A_\zeta\,, \tag{6.27}$$

$$\begin{aligned} &\mathbf{K}^B = \mathbf{K},\quad k_{d-1}^B = \frac{Z_2}{Z_1}\,k_{d-1},\quad k_d^B = \sqrt{\frac{Z_3}{Z_1}}\,k_d, \\ &\Psi_n^B(k^B) = Z_{\Psi_n}^{1/2}\,\Psi_n(k),\quad \phi_\pm^B(k^B) = Z_\phi^{1/2}\,\phi_\pm, \end{aligned} \tag{6.28}$$

and

$$\begin{aligned} &Z_{\Psi_1} = Z_1\left(\frac{Z_1}{Z_2}\right)\sqrt{\frac{Z_1}{Z_3}},\quad Z_{\Psi_2} = Z_4\left(\frac{Z_1}{Z_2}\right)\sqrt{\frac{Z_1}{Z_3}},\quad Z_\phi = Z_7\left(\frac{Z_1}{Z_2}\right)\left(\frac{Z_1}{Z_3}\right)^{3/2}, \\ &\upsilon^B = \frac{Z_5}{Z_4}\left(\frac{Z_1}{Z_2}\right)\upsilon,\quad \kappa^B = \frac{Z_6}{Z_4}\left(\frac{Z_1}{Z_3}\right)\kappa,\quad e^B = Z_e\, e\, \mu^{\frac{\epsilon}{2}},\quad Z_e = \frac{Z_8\sqrt{\frac{Z_1}{Z_2}}}{\left(\frac{Z_1}{Z_3}\right)^{1/4}\sqrt{Z_1\,Z_4\,Z_7}}. \end{aligned} \tag{6.29}$$

A gauge freedom exists that allows us to rescale fields and momenta independently without altering the action. We fix this freedom by imposing $\mathbf{K}^B = \mathbf{K}$, which amounts to measuring the scaling dimensions of all other quantities relative to that of $\mathbf{K}$. The result is the renormalized action $S$—also called the Wilsonian effective action—expressed entirely in terms of renormalized, non-divergent parameters.

### 6.4.1 RG-Flow Equations from One-Loop Diagrams

Gathering our results from the calculations performed at one-loop order, the singular contributions yield

$$Z_1 = 1 - \frac{\mathcal{U}_1\, e^{\frac{4}{3}}}{\epsilon}\, \frac{[\kappa\,(2\,\upsilon - \kappa)]^{\frac{1}{6}}}{\upsilon}, \quad Z_4 = 1 - \frac{\mathcal{U}_1\, e^{\frac{4}{3}}}{\upsilon^{2/3}\,\epsilon},$$
$$Z_2 = Z_3 = Z_5 = Z_6 = Z_7 = Z_8 = 1, \quad \mathcal{U}_1 = \frac{\left[\Gamma(\frac{1}{4})\,\Gamma(\frac{5}{4})\right]^{1/3}}{6 \times 3^{1/6}\,\pi^{4/3}}. \tag{6.30}$$

The fermionic sector is, in general, must be characterised by two independent dynamical critical exponents,

$$z = 1 + \frac{\partial \ln\left(\frac{Z_1}{Z_2}\right)}{\partial \ln \mu}, \quad \tilde{z} = 1 + \frac{\partial \ln\left(\frac{Z_1}{Z_3}\right)}{\partial \ln \mu}. \tag{6.31}$$

Since the one-loop calculation yields $Z_2 = Z_3 = 1$, which implies $\tilde{z} = z$, we treat the two exponents as identical at this order. Anomalous dimensions for the fermionic and bosonic fields are defined through

$$\eta_{\psi_n} = \frac{1}{2}\frac{\partial \ln Z_{\psi_n}}{\partial \ln \mu} \quad \text{and} \quad \eta_\phi = \frac{1}{2}\frac{\partial \ln Z_\phi}{\partial \ln \mu}. \tag{6.32}$$

The beta-functions describing the RG flows of the three coupling constants are

$$\beta_e = \frac{de}{d \ln \mu}, \quad \beta_\upsilon = \frac{d\upsilon}{d \ln \mu}, \quad \beta_\kappa = \frac{d\kappa}{d \ln \mu}. \tag{6.33}$$

The scale $\mu$ entered our formalism purely as a regularization device, introduced to tame the ultraviolet divergences that arise in loop integrals. Since it does not appear in the underlying microscopic theory, physical quantities must be insensitive to its value, and consistency demands that the bare parameters entering $S_{\text{bare}}$ likewise exhibit no $\mu$-dependence. Enforcing this condition, along with the requirement that the finite components of the renormalized quantities take the form of systematic expansions in small $\epsilon$, we must parametrize the above quantities as

$$z = z^{(0)}, \quad \eta_{\psi_n} = \eta_{\psi_n}^{(0)} + \eta_{\psi_n}^{(1)}\,\epsilon, \quad \eta_\phi = \eta_\phi^{(0)} + \eta_\phi^{(1)}\,\epsilon,$$
$$\beta_e = \beta_e^{(0)} + \beta_e^{(1)}\,\epsilon, \quad \beta_\upsilon = \beta_\upsilon^{(0)} + \beta_\upsilon^{(1)}\,\epsilon, \quad \beta_\kappa = \beta_\upsilon^{(0)} + \beta_\kappa^{(1)}\,\epsilon, \tag{6.34}$$

in the limit $\epsilon \to 0$. To arrive at the differential equations governing the RG flow, we proceed through four sequential steps: (1) imposing the condition $\frac{d}{d\ln\mu}$(bare quantity) $= 0$; (2) substituting the expressions from Eqs. (6.30) and (6.34); (3) expanding each equation in powers of $\epsilon$; and (4) matching coefficients of regular powers of $\epsilon$ on both sides to determine all the quantities in Eq. (6.34). Carrying out this procedure yields

$$
\begin{aligned}
&\beta_{\upsilon}^{(1)} = \beta_{\kappa}^{(1)} = \eta_{\psi_n}^{(1)} = \eta_{\phi}^{(1)} = 0, \quad \beta_e^{(1)} = -\frac{e}{2}, \quad z = 1 + \beta_e^{(1)} \frac{\partial Z_1^{(1)}}{\partial e}, \\
&\beta_e^{(0)} = -\frac{e}{4}\left[3(z-1) + e\left(\frac{\partial Z_1^{(1)}}{\partial e} + \frac{\partial Z_4^{(1)}}{\partial e}\right)\right], \quad \beta_{\upsilon}^{(0)} = \upsilon\left(1 - z + \beta_e^{(1)} \frac{\partial Z_4^{(1)}}{\partial e}\right), \\
&\beta_{\kappa}^{(0)} = \kappa\left(1 - z + \beta_e^{(1)} \frac{\partial Z_4^{(1)}}{\partial e}\right), \quad \eta_{\psi_1}^{(0)} = \frac{3(z-1) + 2\,\beta_e^{(1)} \frac{\partial Z_1^{(1)}}{\partial e}}{4}, \\
&\eta_{\psi_2}^{(0)} = \frac{3(z-1) + 2\,\beta_c^{(1)} \frac{\partial Z_4^{(1)}}{\partial e}}{4}, \quad \eta_{\phi}^{(0)} = \frac{5(z-1)}{4}.
\end{aligned}
\tag{6.35}
$$

Solving these equations yields

$$
\begin{aligned}
&z = 1 + \frac{2\mathcal{U}_1\,\tilde{e}\,\kappa^{1/6}(2\,\upsilon - \kappa)^{1/6}}{3\,\upsilon}, \quad \eta_{\psi_1} = \frac{5\mathcal{U}_1\,\tilde{e}\,\kappa^{1/6}(2\,\upsilon - \kappa)^{1/6}}{6\,\upsilon}, \\
&\eta_{\psi_2} = \frac{\mathcal{U}_1\,\tilde{e}\left[2\,\upsilon^{1/3} + 3\,\kappa^{1/6}(2\,\upsilon - \kappa)^{1/6}\right]}{6\,\upsilon}, \quad \eta_{\phi} = \frac{5\mathcal{U}_1\,\tilde{e}\,\kappa^{1/6}(2\,\upsilon - \kappa)^{1/6}}{6\,\upsilon}, \\
&\frac{\beta_e}{e} = \frac{\mathcal{U}_1\,\tilde{e}\left[2\,\upsilon^{1/3} - \kappa^{1/6}(2\,\upsilon - \kappa)^{1/6}\right]}{6\,\upsilon} - \frac{\epsilon}{2}, \\
&\beta_{\upsilon} = \frac{2\mathcal{U}_1\,\tilde{e}\left[\upsilon^{1/3} - \kappa^{1/6}(2\,\upsilon - \kappa)^{1/6}\right]}{3}, \quad \beta_{\kappa} = \frac{2\mathcal{U}_1\,\tilde{e}\,\kappa\left[\upsilon^{1/3} - \kappa^{1/6}(2\,\upsilon - \kappa)^{1/6}\right]}{3\,\upsilon},
\end{aligned}
\tag{6.36}
$$

where $\tilde{e} = e^{4/3}$. To characterize the infrared physics, we track the RG flows with respect to the logarithmic length scale $l$, defined through the derivatives,

$$
\frac{de}{dl} \equiv -\beta_e, \quad \frac{d\upsilon}{dl} \equiv -\beta_{\upsilon}, \quad \frac{d\kappa}{dl} \equiv -\beta_{\kappa}. \tag{6.37}
$$

### *6.4.2 Nature of the Interacting Fixed Points*

We denote the fixed-point values of the coupling constants with a superscript "$*$". For a non-Gaussian fixed point (i.e., one with $\tilde{e} \neq 0$), the last two expressions in Eq. (6.36) require

$$\upsilon^{*1/3} = (\kappa^*)^{1/6}(2\,\upsilon^* - \kappa^*)^{1/6} \quad \Rightarrow \quad \kappa^* = \upsilon^*. \tag{6.38}$$

Inserting this into the condition $\beta_e = 0$ yields

$$\tilde{e}^* = \frac{3\,(\upsilon^*)^{2/3}\,\epsilon}{\mathcal{U}_1}. \tag{6.39}$$

Rather than an isolated fixed point, this defines a fixed line in the three-dimensional space $\{e, \upsilon, \kappa\}$, arising from the continuous family of solutions parametrised by $\kappa^* = \upsilon^*$.

For a trigonally-warped FS, the value of $\kappa$ corresponding to a given $\upsilon$ is extracted from $\epsilon_{\varsigma,\mathbf{p},+}$ [cf. Eq. (6.2)] by varying the warping parameter. Since the FS contour scales as $\left[1 + \frac{\varsigma}{\rho}\cos(3\theta_p)\right]$, one obtains the relations $\upsilon = \frac{\rho-1}{\rho+1}$ and $\kappa = \frac{(\rho-10)(\rho+1)^2}{(\rho-1)^2(\rho+10)}$. A flat patch (with $\kappa = 0$) occurs at $\rho = 10$, implying that for $\upsilon < 9/11 \approx 0.818$, each FS develops a concave region at the hot-spots labeled by superscript "(2)", corresponding to negative $\kappa$. Since our analysis is restricted to convex FS patches, we work exclusively in the regime $\upsilon \gtrsim 0.82$. Comparing the value of $\kappa$ imposed by the trigonal-warping relation with that obtained at the fixed point, we find that the fixed-point value consistently exceeds the noninteracting FS value (see Fig. 6.3). Since $\kappa = 0$ signifies a flat patch, larger $\kappa$ indicates stronger curvature of the FS patch near the hot-spots labeled by superscript "(2)".

To assess the stability of the fixed line, we consider, for a given $\upsilon$, small deviations from the fixed-point values parameterized by $\{\delta\tilde{e}, \delta\kappa\}$. Substituting into the expressions for $\frac{d\tilde{e}}{dl}$ (the negative of the beta function for $\tilde{e}$, obtained from $\frac{de}{dl}$) and

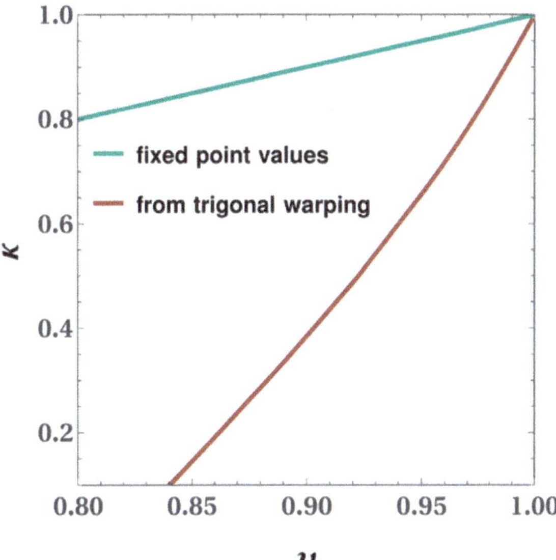


**Fig. 6.3** Illustration of the relationshop between $\upsilon$ and $\kappa$, as dictated by the trigonal-warping dispersion (dark pink trajectory) and the fixed-point requirement $\kappa = \upsilon$ (dark cyan line). The dark pink trajectory is constructed parametrically by varying the warping parameter across the interval $\rho \in [10, 1500]$, with $\upsilon$ and $\kappa$ evolving according to $\upsilon = (\rho - 1)/(\rho + 1)$ and $\kappa = (\rho - 10)(\rho + 1)^2/[(\rho - 1)^2(\rho + 10)]$, respectively

$\frac{d\kappa}{dl}$, and linearizing in the deviation parameters, we construct the stability matrix $\mathcal{M}$ from the coefficients of $\{\delta\tilde{e},\ \delta\kappa\}$ in the two linearized equations. The eigenvalues of $\mathcal{M}$ encode the stability of the fixed point. For the non-Gaussian fixed points, the eigenvalue along the $\tilde{e}$-direction is always negative, while it vanishes along the $\kappa$-direction, indicating that the fixed point is stable with respect to perturbations in $\tilde{e}$ but neutral along the $\kappa$-axis. The same behavior emerges from the stability matrix constructed using $\{\delta\tilde{e},\ \delta\upsilon\}$ when $\kappa$ is held fixed.

## 6.5 RG Flows for $\kappa = \upsilon$

When $\kappa = \upsilon$, the counterterm action assumes the simplified form

$$\begin{aligned} S_{CT} &= \int_k \bar{\Psi}_1(k)\, i \left[A_1\, \mathbf{\Gamma}\cdot\mathbf{K} + \gamma_{d-1} \left(A_2\, k_{d-1} + A_3\, k_d^2\right)\right] \Psi_1(k) \\ &+ \int_k \bar{\Psi}_2(k)\, i \left[A_4\, \mathbf{\Gamma}\cdot\mathbf{K} + \gamma_{d-1}\, A_5\, \upsilon \left(k_{d-1} + k_d^2\right)\right] \Psi_2(k) \\ &+ \frac{1}{2}\int_k A_7\, k_d^2\, \phi(k)\, \phi(-k) - \left[\frac{i\, e\, \mu^{x/2}}{2} \int_k \int_q A_8\, \phi(q)\, \bar{\Psi}_1(k+q)\, \Psi_2(-k) + \text{h.c.}\right], \end{aligned} \tag{6.40}$$

with the one-loop renormalization constants reading

$$Z_1 = 1 - \frac{\mathcal{U}_2\, e^{\frac{4}{3}}}{\upsilon\, \epsilon}, \quad Z_4 = 1 - \frac{\mathcal{U}_2\, e^{\frac{4}{3}}}{\epsilon}, \quad Z_2 = Z_3 = Z_5 = Z_7 = Z_8 = 1, \quad \mathcal{U}_2 = \frac{2^{1/3}}{3^{7/6}}. \tag{6.41}$$

The resulting RG flow equations admit solutions of the form

$$\begin{aligned} z &= 1 + \frac{2\mathcal{U}_2\, \tilde{e}}{3\, \upsilon}, \quad \eta_{\psi_1} = \frac{5\mathcal{U}_2\, \tilde{e}}{6\, \upsilon}, \quad \eta_{\psi_2} = \frac{\mathcal{U}_2\, \tilde{e}\, (3 + 2\, \upsilon)}{6\, \upsilon}, \\ \eta_\phi &= \frac{5\mathcal{U}_2\, \tilde{e}}{6\, \upsilon}, \quad \frac{\beta_e}{e} = \frac{\mathcal{U}_2\, \tilde{e}\, (2\, \upsilon - 1)}{6\, \upsilon} - \frac{\epsilon}{2}, \quad \beta_\upsilon = \frac{2\mathcal{U}_2\, \tilde{e}\, (\upsilon - 1)}{3}. \end{aligned} \tag{6.42}$$

The zeros of the beta functions are located at $\upsilon^* = 1$ and $\tilde{e}^* = 3\, \epsilon/\mathcal{U}_2$. Linearizing about this fixed point, the stability matrix evaluates to $\text{diag}(-\epsilon/2,\ -2\, \epsilon) + \mathcal{O}(\epsilon^2)$. Both eigenvalues are negative, confirming that this is an infrared-stable fixed point in the $\tilde{e}\,\upsilon$-plane. The condition $\kappa = \upsilon = 1$ corresponds to circular FS patches, signaling that the interaction with the CDW boson renormalizes the hot-spots labeled "(2)" until they acquire the same local geometry as those labeled "(1)".

Returning to the more general situation where $\kappa \neq \upsilon$ initially, we found that the infrared fixed points are characterized by $\kappa^* = \upsilon^*$, which forces the RG trajectory into the $\kappa = \upsilon$ subspace irrespective of the starting configuration. In the deep infrared, all hot-spots are thus attracted to the circular-patch fixed point—hence, it collapses to the feature that governs the system in the absence of trigonal-warping.

## 6.6 Conclusion

This chapter has been devoted to investigating a quantum critical point that emerges at the superradiant transition in a cavity QED set-up. Taking as our starting point a honeycomb lattice doped away from half-filling, with low-energy excitations described by Dirac cones, we have identified the CDW wavevectors that connect hot-spots on the resulting FSs, paying careful attention to the role of trigonal-warping in shaping the FS geometry. With this framework in place, we sought to establish rigorously the existence of the NFL phases anticipated by earlier RPA treatments [16]. Employing a systematic perturbative approach grounded in dimensional regularization and RG flow analysis, we demonstrated that infrared-stable NFL fixed points do indeed arise when the RG trajectories are examined along the direction of the coupling constant $e$. The synthetic character of cavity QED platforms affords remarkable flexibility in tuning electronic properties and suppressing unwanted corrections, making them ideally suited for exploring strongly correlated many-body physics through the deliberate manipulation of light-matter coupling. Our results thus contribute to a broader program aimed at establishing the viability of realizing and detecting NFL phases in experimentally accessible cavity QED architectures.

## References

1. F. Schlawin, D. Jaksch, Phys. Rev. Lett. **123**, 133601 (2019). https://doi.org/10.1103/PhysRevLett.123.133601. https://link.aps.org/doi/10.1103/PhysRevLett.123.133601
2. A. Sheikhan, C. Kollath, Phys. Rev. A **99**, 053611 (2019). https://doi.org/10.1103/PhysRevA.99.053611. https://link.aps.org/doi/10.1103/PhysRevA.99.053611
3. J. Li, M. Eckstein, Phys. Rev. Lett. **125**, 217402 (2020). https://doi.org/10.1103/PhysRevLett.125.217402. https://link.aps.org/doi/10.1103/PhysRevLett.125.217402
4. F. Mivehvar, H. Ritsch, F. Piazza, Phys. Rev. Lett. **122**, 113603 (2019). https://doi.org/10.1103/PhysRevLett.122.113603. https://link.aps.org/doi/10.1103/PhysRevLett.122.113603
5. Y. Ashida, A.m.c. İmamoğlu, J. Faist, D. Jaksch, A. Cavalleri, E. Demler, Phys. Rev. X **10**, 041027 (2020). https://doi.org/10.1103/PhysRevX.10.041027. https://link.aps.org/doi/10.1103/PhysRevX.10.041027
6. A. Chiocchetta, D. Kiese, C.P. Zelle, F. Piazza, S. Diehl, Nat. Commun. **12**, 5901 (2021). https://doi.org/10.1038/s41467-021-26076-3
7. A. Chakraborty, F. Piazza, Phys. Rev. Lett. **127**, 177002 (2021). https://doi.org/10.1103/PhysRevLett.127.177002. https://link.aps.org/doi/10.1103/PhysRevLett.127.177002
8. K. Roux, H. Konishi, V. Helson, J.P. Brantut, Nat. Commun. **11**, 2974 (2020). https://doi.org/10.1038/s41467-020-16767-8
9. F. Mivehvar, F. Piazza, T. Donner, H. Ritsch, Adv. Phys. **70**(1), 1 (2021). https://doi.org/10.1080/00018732.2021.1969727

10. X. Zhang, Y. Chen, Z. Wu, J. Wang, J. Fan, S. Deng, H. Wu, Science **373**(6561), 1359 (2021). https://doi.org/10.1126/science.abd4385. https://www.science.org/doi/abs/10.1126/science.abd4385
11. F. Piazza, P. Strack, Phys. Rev. Lett. **112**, 143003 (2014). https://doi.org/10.1103/PhysRevLett.112.143003. https://link.aps.org/doi/10.1103/PhysRevLett.112.143003
12. J. Keeling, M.J. Bhaseen, B.D. Simons, Phys. Rev. Lett. **112**, 143002 (2014). https://doi.org/10.1103/PhysRevLett.112.143002. https://link.aps.org/doi/10.1103/PhysRevLett.112.143002
13. P. Nataf, T. Champel, G. Blatter, D.M. Basko, Phys. Rev. Lett. **123**, 207402 (2019). https://doi.org/10.1103/PhysRevLett.123.207402. https://link.aps.org/doi/10.1103/PhysRevLett.123.207402
14. G.M. Andolina, F.M.D. Pellegrino, V. Giovannetti, A.H. MacDonald, M. Polini, Phys. Rev. B **100**, 121109 (2019). https://doi.org/10.1103/PhysRevB.100.121109. https://link.aps.org/doi/10.1103/PhysRevB.100.121109
15. D. Guerci, P. Simon, C. Mora, Phys. Rev. Lett. **125**, 257604 (2020). https://doi.org/10.1103/PhysRevLett.125.257604. https://link.aps.org/doi/10.1103/PhysRevLett.125.257604
16. P. Rao, F. Piazza, Phys. Rev. Lett. **130**, 083603 (2023). https://doi.org/10.1103/PhysRevLett.130.083603. https://link.aps.org/doi/10.1103/PhysRevLett.130.083603
17. F. Schlawin, D.M. Kennes, M.A. Sentef, Appl. Phys. Rev. **9**(1), 011312 (2022). https://doi.org/10.1063/5.0083825
18. N. Skribanowitz, I.P. Herman, J.C. MacGillivray, M.S. Feld, Phys. Rev. Lett. **30**, 309 (1973). https://doi.org/10.1103/PhysRevLett.30.309. https://link.aps.org/doi/10.1103/PhysRevLett.30.309
19. M. Scheibner, T. Schmidt, L. Worschech, A. Forchel, G. Bacher, T. Passow, D. Hommel, Nat. Phys. **3**(2), 106 (2007). https://doi.org/10.1038/nphys494
20. I. Timothy Noe, G., J.H. Kim, J. Lee, Y. Wang, A.K. Wójcik, S.A. McGill, D.H. Reitze, A.A. Belyanin, J. Kono, Nat. Phys. **8**(3), 219 (2012). https://doi.org/10.1038/nphys2207
21. K. Baumann, C. Guerlin, F. Brennecke, T. Esslinger, Nature **464**(7293), 1301 (2010). https://doi.org/10.1038/nature09009
22. A.H. Castro Neto, F. Guinea, N.M.R. Peres, K.S. Novoselov, A.K. Geim, Rev. Mod. Phys. **81**, 109 (2009). https://doi.org/10.1103/RevModPhys.81.109. https://link.aps.org/doi/10.1103/RevModPhys.81.109
23. P. Sharma, A. Principi, D.L. Maslov, Phys. Rev. B **104**, 045142 (2021).https://doi.org/10.1103/PhysRevB.104.045142. https://link.aps.org/doi/10.1103/PhysRevB.104.045142
24. T. Ando, T. Nakanishi, R. Saito, J. Phys. Soc. Jpn. **67**(8), 2857 (1998). https://doi.org/10.1143/JPSJ.67.2857
25. M.S. Dresselhaus, G. Dresselhaus, Adv. Phys. **51**(1), 1 (2002). https://doi.org/10.1080/00018730110113644
26. Z. Dong, P.A. Lee, L. Levitov, arXiv e-prints (2024)
27. D. Dalidovich, S.S. Lee, Phys. Rev. B **88**, 245106 (2013). https://doi.org/10.1103/PhysRevB.88.245106
28. I. Mandal, Phys. Rev. Res. **2**, 043277 (2020). https://doi.org/10.1103/PhysRevResearch.2.043277. https://link.aps.org/doi/10.1103/PhysRevResearch.2.043277
29. I. Mandal, R.M. Fernandes, Phys. Rev. B **107**, 125142 (2023). https://doi.org/10.1103/PhysRevB.107.125142. https://link.aps.org/doi/10.1103/PhysRevB.107.125142
30. I. Mandal, R.M. Fernandes, Phys. Rev. B **107**, 125142 (2023). https://doi.org/10.1103/PhysRevB.107.125142. https://link.aps.org/doi/10.1103/PhysRevB.107.125142
31. D. Pimenov, I. Mandal, F. Piazza, M. Punk, Phys. Rev. B **98**, 024510 (2018). https://doi.org/10.1103/PhysRevB.98.024510
32. I. Mandal, Nucl. Phys. B **1005**, 116586 (2024). https://doi.org/10.1016/j.nuclphysb.2024.116586
33. I. Mandal, Ann. Phys. **474**, 169925 (2025). https://doi.org/10.1016/j.aop.2025.169925
34. M.A. Metlitski, S. Sachdev, Phys. Rev. B **82**, 075127 (2010). https://doi.org/10.1103/PhysRevB.82.075127

35. M.A. Metlitski, S. Sachdev, Phys. Rev. B **82**, 075128 (2010). https://doi.org/10.1103/PhysRevB.82.075128
36. I. Mandal, Eur. Phys. J. B **89**(12), 278 (2016). https://doi.org/10.1140/epjb/e2016-70509-4
37. G. 't Hooft, Nucl. Phys. B **61**, 455 (1973). https://doi.org/10.1016/0550-3213(73)90376-3
38. S. Weinberg, Phys. Rev. D **8**, 3497 (1973). https://doi.org/10.1103/PhysRevD.8.3497

Zeitfracht Medien GmbH
Ferdinand-Jühlke-Straße 7
99095 Erfurt, Deutschland
produktsicherheit@kolibri360.de